AF718858

WITHDRAWN

Cross-Cultural Risk Perception

Technology, Risk, and Society

An International Series in Risk Analysis

VOLUME 13

Editors

Jeryl Mumpower, *State University of New York, Albany, USA*
Ortwin Renn, *Center of Technology Assessment, Baden-Württemberg, Germany*

The titles published in this series are listed at the end of this volume.

CROSS-CULTURAL RISK PERCEPTION

A Survey of Empirical Studies

Edited by

ORTWIN RENN

Center for Technology Assessment,
Stuttgart, Germany

and

BERND ROHRMANN

University of Melbourne,
Australia

KLUWER ACADEMIC PUBLISHERS

DORDRECHT / BOSTON / LONDON

A C.I.P. Catalogue record for this book is available from the Library of Congress.

HM
1101
.C76
2000

ISBN 0-7923-7747-8

Published by Kluwer Academic Publishers,
P.O. Box 17, 3300 AA Dordrecht, The Netherlands.

Sold and distributed in North, Central and South America
by Kluwer Academic Publishers,
101 Philip Drive, Norwell, MA 02061, U.S.A.

In all other countries, sold and distributed
by Kluwer Academic Publishers,
P.O. Box 322, 3300 AH Dordrecht, The Netherlands.

Printed on acid-free paper

All Rights Reserved
© 2000 Kluwer Academic Publishers
No part of the material protected by this copyright notice may be reproduced or
utilized in any form or by any means, electronic or mechanical,
including photocopying, recording or by any information storage and
retrieval system, without written permission from the copyright owner.

Printed in the Netherlands.

BOOK CONTENTS

PREFACE

Protest against the manifestations of technology has been present since the time of the industrial revolution (Sieferle 1985; Renn 1987). The introduction of trains, steamboats, motorcars, electric lights was always met with skepticism and public discomfort (von Winterfeldt and Edwards 1983). The history books are full of accounts of people's rejection of technological changes. Just to cite one example: In 1824 the daily newspaper of the German city of Augsburg purchased a printing machine driven by a steam engine (Mittelstrass 1998, p. 3). The second editor in chief proclaimed that he would write all his editorials in the nearby open park rather than ever entering the print shop again, another employee terminated his contract with the company and declared his life was endangered. Even pedestrians decided to avoid the street in which the building was located. Although history has recorded numerous examples of unwarranted anxieties, there have been equally worrisome accounts of overconfidence in allegedly fool-proof safety measures and human abilities to cope with disasters. The responses to the change of technology over time seem to oscillate between courage and caution, between overconfidence in the human ability to manage risks and the paralysis of immobility in the light of pending opportunities and threatening hazards.

The patterns of hubris and caution are also present in the recent history of technology and risk. The 1950s were characterized by an almost euphoric belief in scientific and technological progress. It was the time of the "triumphant Prometheus" who provided the promise of unlimited energy, wealth, and comfort. These promises were echoed by the belief in the cornucopia of technological opportunities. This bright vision of the future, however, received its first setback when scientists discovered the ecological boundaries of the economic growth machine. In the late 1960s and early 1970s numerous scientists and activists warned against human hubris and technological overconfidence. Yet most people were convinced that better and more technologies were the answers to the discomforting questions. The idea of unlimited consumption dominated over the concept of precaution. The technical elite was not only able to reassure the public that design criteria for large-scale technologies and risk management practices would be sufficient to contain their catastrophic potential, they were also successful in convincing governments or public management agencies that these technologies had a legitimate role to play in modern society. Risk assessments of the technical elite provided reassurance that the perception of immanent threats was unwarranted. The social psychologist Hans Christian Röglin characterized this attitude as follows: "Modern people love the products of technology, but suppress the process by which these amenities of life are produced" (quoted from Renn 1987, p. 85).

This picture changed dramatically in 1986. This was the year of the Chernobyl catastrophe, the Challenger accident and the pollution of the Rhine river after a fire destroyed a chemical storage building in Basle, Switzerland. These three events had lasting repercussions on public opinion. Even before 1986, many surveys in the United States, Canada, and most of Europe had shown an ambivalent position of most people with respect to the opportunities and risks of large technological systems (Covello 1983; Renn 1983, Slovic 1987; Borcherding et al. 1986, Gould et al. 1988). Risk perception studies and investigations of technological attitudes demonstrated that people were concerned about environmental and health-related impacts of these technologies but, at the same time, assigned a fair proportion of trustworthiness to the technical and political elite. Although some trust had been eroding before 1986, in particular in the aftermath of the accident at the nuclear power plant on Three-Mile-Island, the coincidence of three disasters occurring in the same year caused a major shift in public opinion. How could anyone trust risk professionals after these disasters? The decade between 1986 and 1996 was characterized by a clear defensive attitude of the risk assessment community, a growing distrust in scientific expertise and risk management agencies, and the formation of a powerful counter-elite challenging the official risk assessments of the former experts and demanding new directions in technological policies. Many people voiced the opinion that the risk assessments of the pre-1986 period were discredited by the events of the year 1986.

The technical risk professionals were either hibernating in their own communities or trying (sometimes desperately) to integrate public outrage into their decision or assessment models (Zeckhauser and Viscusi 1996). However, some of the recent conferences on the tenth anniversary of the three disasters witnessed a resurrection of the professional positions on risk (Renn 1998, p. 50). Many technical analysts argued that the so-called disasters of 1986 were not as disastrous as many people had assumed. The 1996 IAEA conference on Chernobyl included a heated debate whether the number of cancer victims exceeded those who had suffered from the curable thyroid cancer, thus bringing the number of victims to a few hundreds at worst. It also seems that the Rhine river accident had hardly any major long-term impact on the river ecosystem (similar, though contested claims have been made for the Alaskan oil spill) and the Challenger disaster appeared to be an exception rather than the rule.

The present situation is full of ambiguities. Most former critics of technological changes have learned that cultural evolution rests on innovation and that innovation implies risk-taking. Opportunities rise out of uncertainties. American sociologist Aaron Wildavsky called the drive towards zero-risk the highest risk of all (Wildavsky 1990). Most technical experts, on the other side, have learnt that good models of risk analysis and stringent methodology in designing technologies can serve only as approximations towards a safer society. Strategies of resilience and flexibility need to accompany safety improvements. Quantitative risk assessments always entail qualitative and subjective elements, such as model specifications or data selection. It can be observed that many political and scientific risk managers advocate a return to the secure haven of numerical risk assessments and question the wisdom of the lay public in judging the seriousness of risks (Okrent 1996). A major problem for both sides has been the philosophical question of whether technical risk estimates represent "objective" probabilities of harm or reflect only conventions of an elite group that may claim no more degree of validity or universal applicability than competing estimates of stakeholder groups or the lay public (Bradbury 1987; Renn 1992, Wynne 1992). For many technical experts, the philosophical position of constructivism seems absurd, for many social scientists and philosophers, the realism of the scientists seem naive at best, and imperialist at worst. Indeed, the concept of a measurable 'real risk' is epistemologically questionable (Brehmer 1994, Rohrmann 1998). At the same time, however, the consequences of risk-taking are real: humans suffer, people die, and the environment is being degraded. Society has an interest to reduce the probability of harm even if the concepts of measuring these probabilities are a construction of the human mind (Rosa 1996).

Will the pendulum swing back to a new era of expert domination in risk policies? The trend seems to go in this direction. At the same time many policy analysts and social scientists warn that ignoring public perception may not only alienate those who feel legitimated to be part of the decision making process and violate democratic principles, but may also underestimate the potential input into the decision making process that the public is capable of providing to the risk manager. If the risk managers fail to find a practical procedure for integrating professional expertise and risk perception, the reemerging conflict between the professional risk assessor and the risk bearer will continue to "haunt" the risk managers and to prevent society to manage risks in accordance with rational criteria of risk reduction and fair burden sharing. Perceptions matter and they should!

This book is an attempt to demonstrate the richness and wealth of theoretical insights and practical information that risk perception studies can offer to policy makers, risk experts, and interested parties. The book begins with an extended introduction summarizing the "state of the art" in risk perception research and provides a brief review of the comparative studies on risk perception in different countries. The main body of the book consists of four major empirical studies focussing on cross-national research:

- Chapter 2 by P. Slovic, J. Flynn, C.K. Mertz, M. Poumadère and C. Mays on risk perception in the United States and France;
- Chapter 3 by B. Rohrmann about a study on hazard perception and evaluation in Australia, Germany, and New Zealand and about a second study on China and Australia;
- Chapter 4 by L. Sjöberg, D. Kolarova, A.-A. Rucai, and M.-L. Bernström on risk attitudes in Bulgaria and Romania (with Sweden as reference); and
- Chapter 5 by E.A. Rosa, N. Matsuda and R.R. Kleinhesselink on the cognitive architecture of risk perception in the United States and Japan.

The last chapter summarizes the results and proposes a framework for understanding and investigating cross-cultural risk perception. This chapter presents the central conclusions of the four studies (and additional literature) and expands the focus on risk perception to issues of risk management and regulation. How should risk perception be included in risk management? Can the results of these comparative studies help decision makers to substantiate their arguments and to improve risk management practice? What are the implications for multi-cultural societies? Finally, lessons for risk communication and further risk perception research complete the last chapter.

We have deliberately chosen studies on cross-cultural comparisons, since the analysis of both common features and distinct particularities help us to distinguish those perception and evaluation processes that reflect more or less universal mechanisms of coping with risk and uncertainty from those which are specific to a culture or a subculture. The main assumption here is that cultural evolution has produced some common cognitive tools and processes that we can detect in almost all cultures of the world while others remain specific responses to the social, cultural and environmental challenges that those cultures face. Furthermore, as Rohrmann (1994) points out, modern societies seem to be divided in many subcultures that have developed their own rationales for perceiving and evaluating risks. These subcultures, however, transcend national or cultural boundaries. Bankers around the world most likely share more values and beliefs with bankers of other cultures or nations than, for example, with their own children. The same can be said for feminists or environmentalists. Globalization has not lead to a unified global culture but rather to a composition of globally operating plural cultures within each society, and these may have similar counterparts in other societies. Consequently, cross-cultural differences can be studied both across countries and within a country, looking at pertinent sub-cultures.

The insights from these comparative studies are crucial for understanding risk perception processes and explaining the often surprising reactions of different people to natural or technological risk. In addition, these insights can be used as "reflexive" knowledge (to pick up a term coined by U. Beck 1992) for policy makers, regulators and risk producers. Knowing what matters to people and comprehending their values and positions is an important precondition to introduce new technologies, to develop risk regulation and to compose risk communication programs. The academic and practical use of these insights for improving our understanding of risk perception and for enhancing our reflective abilities to cope with risk may lead us into a path of a new technological era in which hubris and modesty, courage and caution can be reconciled.

This book would not have been possible without the assistance of many people. We would like to thank Emese Kiss for being a diligent and professional language editor for all chapters. We are grateful to Sigrun Gmelin-Zudrell, Sabine Mücke, Christian D. León, Lidia Ferraro and Dean Lusher for their assistance in preparing the manuscript and helping with layout and proofreading. All the layout work has been performed by Hannelore Zimmermann. We would like to thank her for her enthusiasm and devotion. Last not least we would like to thank our anonymous reviewers and the personnel of the Kluwer Publishing Company.

This publication has been funded by the Center of Technology Assessment in the framework of the project "Risk Evaluation". We are grateful to the Center for its generous support.

Stuttgart and Melbourne
September 1999

Ortwin Renn and Bernd Rohrmann

REFERENCES

Beck, U. (1992), Risk society: Toward a new modernity. London: Sage.

Borcherding, K., Rohrmann, B. & Eppel, T. (1986), A psychological study on the cognitive structure of risk evaluations. In: Brehmer, B., Jungermann, H., Lourens, P. & Sevon, G. (Eds.): New directions in research on decision making, Amsterdam: North- Holland, 245-262.

Bradbury, J. A. (Fall 1989), The policy implications of differing concepts of risk. Science, Technology, and Human Values, 14, 380-399.

Brehmer, B. (1987), The psychology of risk. In: Singleton, W. T. & Hovden, J. (Eds.): Risk and decision, New York: Wiley, 25-39.

Covello, V. T. (1983), The perception of technological risks: A literature review. Technological Forecasting and Social Change, 23, 285-297.

Gould, L. C., Gardner, G. T., DeLuca, D. R., Tiemann A, Doob, L. W., and Stolwijk, J. A. J. (1988), Perceptions of technological risks and benefits. New York: Russel Sage Foundation.

Mittelstrass, J. (1998), Zwischen Prometheus und Kassandra. Licht und Dunkel in der Welt des technischen Verstandes. Lecture at the Forum for Young Scientists at the University of Wittenberg/Halle. Manuscript from the Department of Philosophy. University of Konstanz.

Okrent, D. (1996), Risk perception research program and applications: Have they received enough peer review? In: C. Cacciabue and I. A. Papazoglou (eds.): Probabilistic safety assessment and management '96 ESREL'96-PSAM'96. Berlin: Springer, pp. 1255-1259.

Renn, O. (1983), Technology, risk, and public perception. Applied Systems Analysis, 4, 50-65.

Renn, O. (1987), Eine kulturhistorische Betrachtung des technischen Fortschritts. In: H. Lübbe (ed.): Fortschritt der Technik - gesellschaftliche und ökonomische Auswirkungen. Heidelberg. R. von Deckers Verlag, pp. 65-100.

Renn O. (1992), Concepts of risk: a classification. In: S. Krimsky and D. Golding (eds), Social theories of risk. Westport, CT: Praeger, pp. 53-79.

Renn, O. (1998), The role of risk perception for risk management. Reliability Engineering and Systems Safety, 59, 49-62.

Rohrmann, B. (1994), Risk perception of different societal groups: Australian findings and cross-national comparisons. Australian Journal of Psychology, 46, 150-163.

Rohrmann, B. (1998), The risk notion - epistemological and empirical considerations, in: Stewart, M.G., & Melchers, R.E. (Eds.): Integrative risk assessment; Rotterdam: Balkema, 1998 pp 39-46.

Rosa, E. A. (1996), Metatheoretical foundations for post-normal risk. Journal of Risk Research, 1 (Vol. 1), 49-72

Sieferle, R. P. (1985), Fortschrittsfeinde? Opposition gegen Technik und Industrie von der Romantik bis zur Gegenwart. München. Beck

Slovic, P. (1987), Perception of risk. Science, 236, 280-285.

von Winterfeldt, D. and Edwards, W. (1983), Patterns of conflict about risky technologies. Research Paper SSRC-1-12-83. Los Angeles. University of Southern California

Wildavsky, A. (1990), No risk is the highest risk of all. In: Glickman, T.S. and Gough, M. (eds.): Readings in risk. Washington, D.C. Ressources for the Future, 120-127

Wynne, B. (1992), Risk and social learning: Reification to engagement, in: S. Krimsky and D. Golding (eds.), Social theories of risk, Westport: Praeger, pp. 275-297

Zeckhauser, R. and Viscusi, K. W. (1996), The risk management dilemma. Annals of the American Academy of Political and Social Science, Special Issue, H. Kunreuther and P. Slovic (eds.), Challenges in risk assessment and risk management. Thousand Oaks. Sage, 144-155

1

RISK PERCEPTION RESEARCH

AN INTRODUCTION

Bernd Rohrmann
University of Melbourne, Melbourne, Australia

Ortwin Renn
Center of Technology Assessment, Stuttgart, Germany

TABLE OF CONTENTS

SUMMARY

This first chapter provides a brief review of the present knowledge on risk perception and elucidates the psychological and social factors that shape the experience of risk. It aims at integrating the results of psychological, sociological, and cultural studies and presents the major findings of the social sciences as they seem relevant for a deeper understanding of risk and the role of risk perception for risk management. A second objective is to review comparative studies about risk perception in different cultures and nations and the approaches utilized in such research.

After discussing definitions of risk, two sections introduce the methodology of cross-cultural risk research. Two further sections deal with core substantive issues, including heuristics, hazard features, psychological facets of hazard appraisals, pertinent group processes and societal factors determining risk perception. Finally, the relevance and utilization of risk perception research for the public discourse about hazards and implications for managing risks are outlined.

1 THE CONCEPT OF RISK

Talking about risks faces the immediate danger that everybody talks about something different. In fact, the understanding of the concept "*risk*" differs widely across sciences and scientists (see, e.g., Cohrssen & Covello 1989, Drottz-Sjöberg 1991, Fischhoff, Watson & Hope 1984, Renn 1992a, Rohrmann 1998, Short 1984, Vlek 1996, Yates & Stone 1992). There is no commonly accepted definition for the term risk - neither in the sciences nor in public understanding. In disciplines such as engineering, physics, pharmacology, toxicology or epidemiology, formal definitions based on the probability and physical measurements or corresponding utilities of negative outcomes are preferred; quantification of probabilities and outcomes lie at the core of this approach. In the social sciences, the 'meaning' of risk is a key issue, and qualitative aspects of risk are seen as crucial facets of the concept.

Most risk concepts have one element in common, however: the distinction between reality and possibility (Markowitz 1991; Evers and Nowotny 1987). If the future were either predetermined or independent of present human activities, the term "risk" would make no sense. If the distinction between reality and possibility is acknowledged, the term "risk" is often associated with the possibility that an undesirable state of reality (adverse effects) may occur as a result of natural events or human activities (similar in Kasperson and Kasperson 1987: viii; Starr and Whipple 1991: 53; Hillgartner 1992: 40; the same idea is expressed in more technical terms in: National Research Council 1983; IEC 1993). It may be difficult to determine, however, what characteristics are necessary to label an outcome as "adverse" rather than "desirable" or "tolerable" (cf. discussion in: Renn 1990a; HMSO 1988). In particular, if one intends to include the common use of the term "risk" in economic theory, both gains and losses need to be subsumed under "risk". Similarly, in decision theory "risk" basically stands for uncertainty.

In addition, Machlis and Rosa pointed out that there is a phenomenon such as "desired risk" (for example in some sports or "adventure" activities) that people aspire to reach for experiencing a special thrill (Brengelmann 1991, Horvath & Zuckerman 1992, Machlis and Rosa 1990). Riskiness then gets a positive rather than a negative connotation. Rosa hence recommended to use the term "risk" for uncertain outcomes regardless whether they are positive or negative (Rosa 1996).

Given such arguments, risk can be understood as *the possibility that human actions, situations or events might lead to consequences that affect aspects of what humans value*. Such a definition implies that humans can and will make causal connections between actions (or events). Consequences are perceived from a non-fatalistic viewpoint. They can be altered either by modifying the initiating activity or event or by mitigating the impacts (Appelbaum 1977). Risk is therefore both a descriptive and a normative concept. It includes the analysis of cause-effect relationships, which may be scientific, anecdotal, religious or magic (Douglas 1966; Wiedemann 1993), but it also carries the implicit message to reduce undesirable effects through appropriate modification of the causes or, though less desirable, mitigation of the consequences.

However, in most contexts "risk" refers to a danger of unwanted events, rather than an opportunity for desired outcomes. Then *risk* can be defined as *the possibility of physical or social or financial harm/detriment/loss due to a hazard within a particular time frame*. "Hazard" refers to a situation, event or substance that can become harmful for people, nature or human-made facilities. "People" at risk might be residents, employees in the workplace, consumers of potentially hazardous products, travelers/commuters and/or the society at large.

After defining the term risk, it is important to look at activities that are associated with the analysis of risk from a scientific as well as a layperson's perspective. Let us first turn to the term "risk assessment". *Risk assessment* denotes the scientific process of defining the components and implications of a hazard in precise, usually quantitative terms. In technical risk assessments (e.g.: PRA = Probabilistic Risk Analysis), this means specifying what is at stake, calculating the probabilities for (un)wanted consequences, and aggregating both components, usually by multiplying the probabilities by the magnitude of the effects (cf. e.g. Covello & Merkhofer 1993, Kolluru and Brooks 1995). *Risk management* refers to the process of reducing risks (either the hazard itself or its consequences) to a level deemed tolerable by society and to assure control, monitoring, and public communication (Morgan 1990, Kolluru 1995; Fischhoff 1996). Since risk refers to a potential of "real" consequences, it is both a social construction and a representation of reality (Renn 1992a; Renn et al. 1992; Short 1989). This implies, risk is not a physical entity but an inference about implications of a current or future reality.

Within the social sciences, people's views on risks are usually labeled *risk perception*. Principally, risks cannot be "perceived" (as Brehmer 1994, p. 83 stated: "to speak of 'perceived risk' in the same manner we speak of 'perceived length' makes no sense"). However, this terminology has become the standard convention in the scientific community. People receive and process physical signals (such as witnessing an explosion), as well as information about possible outcomes of (future) human actions or natural events, and they tend to form respective opinions and attitudes towards risk sources and impacts. The term "risk perception" refers to people's judgments and evaluations of hazards they (or their facilities, or the environment) are or might be exposed to. Both experiences and beliefs are to be considered.

The use of the risk terminology differs considerably in different disciplines. The understanding of "risk" in natural sciences tends to clash with concepts of risk in the social sciences (cf. Rohrmann 1998). For example, quite often the term "real" or "actual" risk is used as counterpart to "perceived risk" - which is scientifically inadequate, psychologically derogative and not helpful for the communication between disciplines. Epistemiologists could easily show that all statements about risks, whether rough guesses or highly quantitative data-based computations, are only reflections of the "reality" under consideration (see fig. 1 for an illustration). It appears more appropriate to label results from quantitative risk assessments (which can be seen as a model-based estimate of the "real" risk) as "statistical" or "statistically estimated" or "probabilistic" or "predicted" risk - which then may be contrasted to perceived risk.

Figure 1. **ISSUES OF RISK PERCEPTION RESEARCH**

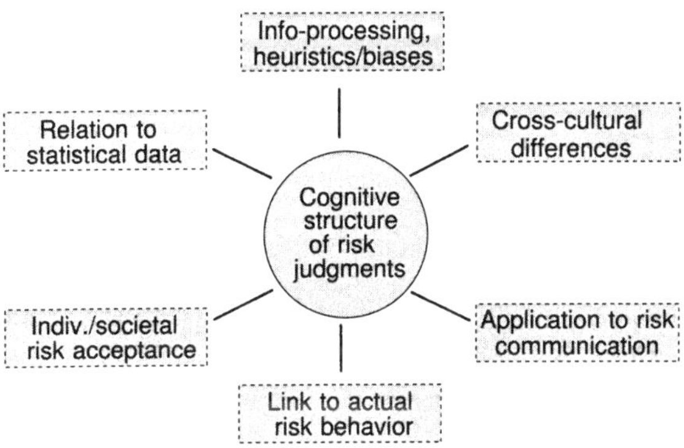

The constructive nature of the term risk should not be misunderstood, however, as if the actual damage that can occur as a result of risk-taking behavior would constitute only a "mental construct" without correspondence to real losses. Damages and losses are always real but the concepts to predict the likelihood of such events depend on mental models constructed in the realm of human knowledge and beliefs.

Some authors actually mean perceived risk *level or magnitude* when talking about "perceived risk"; however, generally the label risk perception is meant to include all aspects of rating and evaluating risks. Risk perception research has several facets (cf. fig. 2): The analysis of risk judgments (the core interest) is usually extended to factors of risk acceptance (in individual or societal terms). Psychologists are particularly interested in the underlying information processes as well as the link to actual behavior in risk situations. Furthermore the findings can be related to statistical hazard data and are substantial for risk communication programs. Recently cultural differences in risk perception have emerged as an important topic. In each field, different approaches and methodologies are pursued in both basic and applied research. The following sections will describe the various approaches to risk perception research and provide a brief review of the main results pertaining to psychological, sociological and cultural studies.

Figure 2. **PERCEIVED, "REAL", MODELLED RISK**

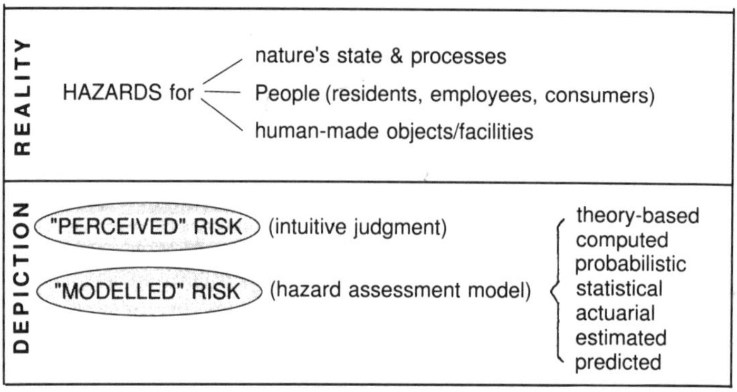

2 APPROACHES TO RISK PERCEPTION

2.1 Risk research in the social sciences

"Risk" has been a highly topical term for more than a decade now. Risks related to technological developments, working conditions, residential settings, private activities, public health, environmental hazards, global ecological changes and so on have become prominent issues of political/ societal discourse, as well as of academic and applied research. Salient examples are car accidents, asbestos, smoking, drugs, AIDS, nuclear energy, water pollution, genetic engineering, ozone layer depletion. Some disasters, such as the accidents in Bophal or Chernobyl, the earthquakes in San Francisco, Kobe or recently in Turkey, the oil catastrophe in Kuwait, or the bushfires in Australia, have further increased the awareness of risks threatening humankind (Fischer, Morgan, Fischhoff, Nair & Lave 1991, MacGregor 1991). Particularly technology-induced hazards get substantial publicity and are widely discussed in society.

Risk issues also create controversies. In many societies, severe conflicts about the evaluation of risks have emerged, especially with respect to large-scale technologies such as chemical industries, nuclear energy and genetic engineering (Bechmann 1993, Beck 1986/1992, Brown 1988, Cvetkovich & Earle 1992, Jungermann, Rohrmann & Wiedemann 1991, Sjöberg 1987, Vlek & Keren 1992, v. Winterfeldt & Edwards 1984). Two concerns are relevant here: risks to human health and well-being; and risks to the state of the environment, emerging from the impact of human activities. Impacts may be acute, as in the case of an accident or disaster; or chronic, in the case when a hazardous process continues over time.

As manifold effects are to be considered, the assessment of such risks is a very complex task. Furthermore, there is a considerable gap between how experts think about risks and how non-professional people judge and evaluate risks. Depending on the underlying definitions and criteria, very heterogeneous risk assessments are given by different groups (Edwards & v. Winterfeldt 1987, Fischhoff, Slovic & Lichtenstein 1981, Kemp 1993). Many of the risks most prominent in the view of the public are not those which - according to statistical data - actually result in high accident figures, mortality rates, health impacts, and so on. Apparently many more factors than those data influence risk perception, risk behavior and risk management, including a variety of social, psychological, ethical aspects (O'Riordan 1983). Furthermore, cultural factors are crucial for people's risk evaluation. In fact, "each society has its own distinctive portfolio of risks that are believed to be of concern" (Johnson & Covello 1987). Thus societal decision making about risk issues has become more and more complicated. This is particularly true for hazards which have both physical and socio-psychological impacts.

Under these circumstances 'technical' risk analysis (as done in natural sciences or economics) is increasingly complemented by social-science approaches in order to expand how "risk" is conceptualized and to understand how humans experience, rate and evaluate risks they are (or might be) exposed to (Brehmer 1987, Handmer, Dutton, Guerin & Smithson 1991, Heimer 1988, Rayner & Cantor 1987, Renn 1983, Vlek & Stallen 1980, Waterstone 1991). The social science perspectives on risk broaden the scope of undesirable effects, include other ways to express possibilities and likelihood, expand the understanding of "perceived" reality and help to explain how individuals and societies at large "socially construct" their view on undesirable realities (Bradbury 1989; Brehmer 1987, O'Riordan 1983, Renn 1992a).

The following sections will provide a review of the main approaches to study and explain risk perception. In a first phase, pertinent studies were interested in general principles of risk perception and appraisal. More recently, the focus is on cultural differences within and across societies.

2.2 The psychometric paradigm

General mechanisms of people to process risk information have been intensively studied, predominantly by psychometric quantitative methods. The so-called psychometric approach is based on four intentions:

- to establish "risk" as a subjective concept, not an objective entity,
- to include technical/physical and social/psychological aspects in risk criteria,
- to accept opinions of "the public" (i.e., laypeople, not experts) as the matter of interest,
- to analyze the cognitive structure of risk judgments, usually employing multivariate statistical procedures such as factor analysis, multi-dimensional scaling or multiple regression.

This approach to risk research was originated in the work of B. Fischhoff, S. Lichtenstein and P. Slovic, the "Oregon Group" (see Fischhoff, Slovic & Lichtenstein, Read & Combs 1978, Lichtenstein, Slovic, Fischhoff & Combs 1978, Slovic, Fischhoff & Lichtenstein 1980; cf. also Slovic 1992). Quite a number of studies followed, most of them in the USA and European countries. For a comprehensive review and documentation of this body of research see Rohrmann 1995/1999; overviews are provided by Fischhoff, Bostrom & Quadrel 1993, Guerin 1991, Jungermann & Slovic 1993, Pidgeon, Hood, Jones, Turner & Gibson 1992, Renn 1986, 1990;

methodological issues are discussed in Arabie & Maschmeyer 1988, Hartenian, Bobko & Berger 1993, Rohrmann 1995, Slovic 1992.

The subjective meaning of risk concepts, evaluation of risk sources and determinants of risk acceptance have also been investigated by means of qualitative approaches (such as exploratory interviewing, focus groups, repertory-grid procedures and content analysis). Some of these studies include Earle & Lindell 1984, Fischer, Morgan, Fischhoff, Nair & Lave 1991, MacGregor 1991, Perusse 1980, Tyszka & Goszczynska 1993, Wiedemann & Kresser 1997.

2.3 Sociological and "cultural" approaches

Empirical (mostly psychological) studies on social and cultural factors in risk perception are predominantly based on individual or group responses to risk issues. However, the process of risk perception in society has been thoroughly analyzed from a 'macro-sociological' perspective as well - see for example Beck 1992, Dake 1992, Douglas & Wildavsky 1982, Heimer 1988, Luhmann 1990, 1993, Renn 1992b, Schuez 1990, Thompson 1980a, Wildavsky 1988, 1993 (most of this work is theoretical; empirical research is fairly rare).

Sociologists have particularly stressed that the evaluative process of risk perception is determined by the norms, value systems and cultural idiosyncrasies of societies (see, e.g., Beck 1992, Cvetkovich & Earle 1991, Dake 1991, Douglas & Wildavsky 1982, Rayner 1992, Sjöberg 1995, Thompson, Ellis & Wildavsky 1990). According to the "cultural approach", risk is a "social and cultural construction" (Johnson & Covello 1987) - not an 'objective' entity to be measured independently of the context in which hazards occur. Consequently each society or social group may have its own specific set of risks believed to be of concern, based on its prevalent "worldview". Within this area, research has been more often qualitative than quantitative, including philosophical treatises of risk perception.

Empirically, cultural differences in risk perception can be investigated from several perspectives, based on intra-national group comparisons or cross-national studies; (cf. Table 1 for related issues). Most of the research conducted so far follows the psychometric paradigm. Some researchers (e.g., Marris, Longford & O'Riordan 1996) have attempted to bridge the gap between psychological and sociological conceptualizations of risk perception research.

Table 1. **NOTIONS OF „CROSS-CULTURAL" RISK PERCEPTION RESEARCH**

Level of comparison	intra-national	inter-national
Units of study	professional or ideological subgroups of society	countries or cultures
Core variables	beliefs & attitudes towards perceived risk sources	culturally embedded values regarding safety & risk

3 FACETS OF EMPIRICAL RISK PERCEPTION STUDIES

The many empirical studies published over the last 15 years differ considerably in both substantial and methodological respects. Thus it might be helpful to identify the main facets of this kind of research. Some distinctions will be briefly discussed. In principle, a study on risk perception constitutes a 3-dimensional data cube related to hazards (risk sources), judgmental aspects and characteristics of respondents. Studies diverge widely with respect to these facets, as well as in their psychometric approach and type of data analysis.

3.1 Hazards (risk sources)

Most studies deal with a multiple and (deliberately) heterogeneous set of hazards. Some concentrate on a particular sort of hazards (e.g., technologies, activities, substances, health issues, etc.), and others focus on one specific risk (most frequent issues are large-scale technologies, above all nuclear energy, but also some diseases, e.g. AIDS and smoking).

A basic taxonomy of risk sources with respect to the subject of risk, types of personal risk exposure and kinds of effects is presented in Table 2. It should be noted that most studies deal with two sorts of risks: those to which the respondents are actually exposed (e.g., judgments of traffic hazards by car drivers or judgments of nuclear power generation by local residents), and risks of which the respondents have no knowledge based on personal experience (the latter case is much more frequent).

Table 2. **TAXONOMY OF RISK SOURCES**

Subject of risk:
- Risks for the state of the environment
- Risks for human's health, well-being and their assets

Types of personal risk exposure:
- Individual activities
  - occupational
  - private
- Residential conditions
  - natural hazards
  - technology-induced hazards

Kind of effects:
- physical / financial / social
- acute / chronic
- local / regional / global
- present / next generation / future

3.2 Judgments of risk aspects and dimensions

Usually the selected risk sources are to be judged by respondents according to a set of predefined characteristics. The core variables in risk perception research are the (perceived) magnitude of the risk and risk acceptance. However, in most studies many more risk-related aspects have been employed (see Table 3). Table 3 lists examples for five types of variables: Risk level aspects, qualitative features of hazards (comprising a variety of factors which increase or decrease the perceived riskiness and acceptability of a hazard), benefit aspects, personal relation to the hazard, and acceptability aspects. Some studies deliberately avoid to define risk criteria and just ask for the similarity of risk sources, e.g., in pair comparisons. In that case risk dimensions are considered as an output to be searched for in the investigation, not as an input into the study.

So far, the outlined risk aspects have been predominantly investigated in quantitative terms; however, qualitative approaches have been employed as well (e.g. Earle & Lindell 1984, MacGregor 1991 Wiedemann & Kresser 1997) and seem to gain increased attention in recent years.

3.3 Types of respondents and sampling

In most studies, the sampling of respondents is based on ad-hoc groups (usually students or other convenience sampling) rather than representing specified target populations. In a number of studies, person-related variables such as age, sex, occupation, political and ecological attitudes, personality characteristics and so on have been considered in order to analyze their influence on risk evaluation. Also, specific societal groups (e.g., teachers, 'green' organizations, company employees) have been investigated. The more cultural (intra-national and cross-national) differences in risk perception became obvious the less acceptable ad-hoc sampling was. Sample sizes are predominantly small, and only few studies can claim to provide a nation-wide representative sampling.

3.4 Countries and cultures

Within the last decade, quite a number of cross-national studies have been conducted. There are two approaches: Several researchers replicated the Slovic/Fischhoff/Lichtenstein approach and compared their findings 'posthoc' with USA data, while others conducted genuine two-country research. In Table 4 the setup of 14 cross-national studies is detailed (for a comprehensive documentation see Rohrmann 1999). Further studies deal with one risk source only, namely nuclear power, e.g., Eiser et al. 1990 (England/France/Germany/Holland/ Australia) or Renn & Swaton 1984 and Swaton & Renn 1984 (Germany, Japan, Philippines). These studies aim at cultural differences, mostly across European and North American (i.e., "western") countries. Recently there has been an increased interest in Asian countries such as Japan and China.

Researchers have also looked at cultural (sub-)groups within a country, comparing the risk perception of people adhering to a particular 'worldview' (cf. section 1.4 above); examples are Borcherding, Rohrmann & Eppel 1986, Marris et al. 1996, Rohrmann 1994, Wildavsky & Dake 1990. Up to now there seems to be no risk perception research in African countries, and Asian or South American studies are still scarce.

Table 3. **RISK-RELATED VARIABLES - RISK ASPECTS JUDGED BY RESPONDENTS**

Risk level aspects

> Magnitude of risk; Overall risk rating
> Estimated number of fatalities/year
> (Individual) Fatality probability; Loss of life expectancy
> Danger of health impacts/harm
> (Un)Avoidability of accidents/disasters
> Catastrophic potential; Number of fatalities in a disaster

Qualitative features of hazards

> Dreadfulness; Fear associations; Fright
> (Un)Familiarity; Known to exposed; Imaginability of dangers
> (Not) Observable (source a/o effects)
> (Un)Controllability
> (Un)Known to science
> Immediate vs. delayed impacts
> Impacts for future generations

Benefit aspects

> Benefits provided by risk source for an individual
> Attractiveness of a risky activity
> Societal benefits, contribution to human needs
> Relevance for human needs
> Risks and benefits equitable

Personal relation to hazard

> (In)Voluntarily exposed; Personal influence on choice
> Degree of concern; Being worried
> Personal exposure; Personally affected
> (Individual) Risk-reducing involvement; Action taken
> Desired distance to hazardous facility

Acceptability aspects

> Willingness-to-pay (for risk mitigation)
> Desired level of restriction
> Risk acceptance: individual perspective
> Societal acceptability of a risk

Table 4. **SELECTED CROSS-CULTURAL RISK PERCEPTION STUDIES**.

STUDY	*Countries*	Sample Size(s)	Sub-groups	Risk Sources	Risk Aspects
WINTERFELDT et al. 1984	*USA+Germany*	57+68 *	-	14	4
ENGLANDER et al. '86	*Hungary* *(USA)*	30/29 * (175)	-	90/30	1/9
TEIGEN et al. '88	*Norway* *(USA)*	37/35/64 * (175)	-	30/90/35	9/1/9
HOEFER / RAJU '89	*France/USA*	50+26 *	-	6	10/1
KEOWN '89	*HongKong* *(USA)*	65 * (175)	-	30/15	2/6
MECHITOV / REBRIK '90	*Russia* *(USA)*	24/24 (175)	2	13/9/75	4/7/1
KLEINHESSELINK / ROSA '91	*USA/Japan*	62+69 *	-	70	7
GOSZCZYNSKA et al. '91	*Poland* *(USA)*	140 (175)	4	40	1/15
ROHRMANN '89/93 + BORCHERD. & R.'86	*Germany+NZ+ Australia*	217+278+263	4/8	24	11
HINMAN et al. '93	*Japan/USA*	290+747	-	30	4
NYLAND 1993	*Braz./Swed.*	144+119	5/4	100/33	1/1
SOKOLOWSKA / TYSZKA '95	*Poland/ Sweden*	601+1488	9	6+11	2
SLOVIC et al. *this vol.*	*France/USA*	1550+1512	-	25	5
SJOEBERG et al. *this vol.*	*Bulg./Rom.*	240+192	6	100/28	7
ROSA et al. *this vol.*	*Japan/USA*	273+152 (*)	-	70	7
ROHRMANN & CHEN '99	*Australia/ China*	203+270 (*)	4	25	10

Notes: „(USA)“ refers to secondary comparisons with USA data published by Fischhoff et al. 1978/1985.
„#/#“ indicates sub-samples within a study. * = student samples.

3.5 Modes of data analyses

Statistical analyses of risk perception data mainly focus on two issues: Mean differences in judgments with regard to risk sources and/or groups of respondents (usually analyzed by (M)ANOVA procedures); and the cognitive structure of risk aspects, employing multivariate methods such as factor analysis, multiple regression, and structural/causal modeling (those analyses are to be based on the inter-correlations among the substantial risk-related variables).

A note on modes of data analysis in the case of 3-dimensional data decks (i.e., x judgments on y risk sources by n respondents): Correlation among risk aspects can be determined in several ways:

- *across risk sources:* the respective correlation matrix could be computed for each respondent or for mean data decks (either means of the whole sample or means of specific subgroups);
- *across respondents:* an analogous matrix could be computed for each risk source or mean ratings based on defined sets of risks.

Unfortunately the chosen method is often not explicated even though these approaches will yield quite different results (see e.g., Borcherding, Rohrmann & Eppel 1986, Johnson & Tversky 1984, Slovic 1992); therefore a careful interpretation is indicated. This is particularly true with respect to the use of aggregate data (cf. e.g., Marris et al. 1996, Rohrmann 1995, Schuetz in prep., Sjöberg 1996), as mean data tend to produce deceptively strong contingencies which might not hold for analyses based on individual data. There is also a danger that excluding inter-individual variance increases the influence of semantic similarities among risk aspects on factor-analytic results. Both differentiated sampling and disaggregate data analyses are necessary to reflect the subjective risk evaluation process.

A beneficial result of the described diversity (regarding hazards, aspects, sampling and analyses) is the rich body of knowledge about risk perception. At the same time, however, the restricted comparability and cross-validation can be considered as a disadvantage. In any case it is indispensable to check carefully a study's methodological approach before findings are interpreted and generalized.

4 CORE FINDINGS ON THE STRUCTURE OF RISK PERCEPTION

The available social-science research on the meaning of risk, the subjective understanding and evaluation of risk sources and the determinants of risk acceptance is manifold and heterogeneous. Not surprisingly, results depend strongly on the type of hazards under study, the substantive risk criteria, the characteristics of the surveyed respondents, and also on the statistical methods used in the data analyses. Thus it might appear somewhat futile to look for a general synopsis of findings - in fact there isn't and can't be 'the main finding' in this field. However, most researchers agree that particular topics need to be analyzed systematically in order to improve the understanding of human risk perception.

In parts 4 and 5 of this introductory chapter, selected findings shall be summarized with regard to three questions: How do individuals perceive and judge risk? How do social factors influence risk perception? And how are the various facets of perceiving risks linked?

4.1 Influence of attention and selection

Most risks that modern society faces are not experienced by human senses but learned through communication. Rarely do modern humans face disasters personally; however, the media provide them with ample information about hazardous events wherever they take place. The dangers of technologies or nature, the risks of food additives or chemicals in drinking water, the threat of nuclear disaster or a chemical explosion would probably never reach public attention unless society communicates about these adverse possibilities. Risk perception is less a product of experience or personal evidence than a result of social communication (Luhmann 1986).

This observation has major consequences: Today's society provides an abundance of information, much more than any individual can digest. It is assumed that the average person processes only a tiny fraction of all the information to which he or she is exposed. Most information is simply ignored. This is not a malicious act but a sheer necessity in order to reduce the amount of information a person can process in a given time. Societal evolution has provided humans with an almost automated and often subconscious tool of selecting the important information from the abundance of information supplies.

In order to economize information processing, individuals are likely to evaluate whether it is necessary to study the content of the information in detail or to make a fast judgment according to some salient cues in the message received. The first strategy refers to the central route of information processing, the second to the peripheral route (Petty and Cacioppo 1986; Renn and Levine 1991). In the central mode, the receiver performs two types of evaluations: first, an assessment of the probability that each argument is true; and second, an assignment of weight to each argument according to the personal salience of the argument's content. The credibility of each argument can be tested by referring to personal experience, plausibility, and perceived motives of the communicator. The major incentives for changing an attitude in the central mode are the proximity with and the affinity to one' own interests, values, and world views.

In the peripheral mode, receivers do not bother to deal with each argument separately, but look for easily accessible clues to make their judgment on the whole package. Examples of such cues are the length of a message, the number of arguments, the package (color, paper, graphic appeal, and others), and the presence of symbolic signals that trigger immediate emotional responses (cf. Kasperson et al. 1988).

4.2 Intuitive heuristics

Once information has been received, common sense mechanisms process the information and help the receiver to draw inferences. These processes are called intuitive heuristics. They are particularly important for risk perception since they relate to the mechanisms of processing probabilistic information. Early psychological studies focused on personal preferences for probabilities and attempted to explain why individuals do not base their risk judgments on expected values, i.e. the product of probability and magnitude of an adverse effect (Pollatsek and Tversky 1970; Lopes 1983). One of the interesting results of these investigations was the discovery of systematic patterns of probabilistic reasoning that are well suited for most everyday situations. People are risk averse if the stakes of losses are high and risk prone if the stakes for gains are high (Kahneman and Tversky 1979). Many people balance their risk taking behavior by pursuing an optimal risk strategy which does not maximize their benefits but assures a satisfactory payoff and the avoidance of major disasters (Luce and Weber 1986).

Second, more specific studies on the perception of probabilities in decision making identified several biases in people's ability to draw inferences from probabilistic information (Festinger 1957; Kahneman and Tversky 1979; Ross 1977; Renn 1990a). These biases are summarized in Table 5.

Table 5. **INTUITIVE BIASES OF RISK PERCEPTION**

BIASES	DESCRIPTION
Availability:	Events that come to people's mind immediately are rated as more probable than events that are less mentally available.
Anchoring effect:	Probabilities are adjusted to the information available or the perceived significance of the information.
Representativeness:	Singular events experienced in person or associated with properties of an event are regarded as more typical than information based on frequencies.
Avoidance of cognitive dissonance	Information which challenges those perceived probabilities that are an integral part of an existing belief system will either be ignored or downplayed.

Although these biases constitute clear violations of logical rules, they might have been overrated in the literature (Fischhoff et al. 1981). Many laboratory situations provide insufficient contextual information to provide enough cues for people on which they can base their judgments (Lopes 1983). Relying on predominantly numerical information and being unfamiliar with the subject, many subjects in these experiments retrieve to 'rules of thumb' in drawing inferences. In many real life situations, experience of and familiarity with the context provide additional information to calibrate individual judgments, particularly for nontrivial decisions (cf. Heimer 1988). A useful wider perspective is to study "mental models" which humans develop for hazards they encounter (Bostrom 1990, Bostrom et al 1992, Jungermann et al. 1988.) Nevertheless, risk managers should be aware of these biases because they are found in public perception and may be one of the underlying causes for the observed public response.

4.3 Qualitative aspects

Attributes of risk sources, the type of risk involved and its situational characteristics, as well as characteristics of hazard outcomes shape individual risk estimations and evaluations. Various psychometric methods have been employed to explore these qualitative characteristics of risks. The following contextual variables of risk have been found to affect the perceived seriousness of risks (Slovic et al. 1981a; Vlek and Stallen 1981; Covello 1983; Borcherding et al. 1986, Gould et al. 1988; Renn 1990a; Jungermann and Slovic 1993, Rohrmann 1994):

- *the expected number of fatalities or losses:* Although the perceived average number of fatalities correlates with the perceived risk of a technology or activity, the relationship is weak and generally explains less than 20 percent of the declared variance.
- *the catastrophic potential:* Most people show distinctive preferences among choices with identical expected values (average risk). Low-probability high-consequence risks are usually perceived as more threatening than more probable risks with low or medium consequences; they tend to create fear and repulsion.
- *the perceived properties of the risk source or the risk situation:* Among the most influential factors are: the perception of dread with respect to the possible consequences; the conviction of having personal control over the magnitude or probability of the risk; the familiarity with the risk; the perception of equitable sharing of both benefits and risks; and the potential to blame a person or institution responsible for the creation of a risky situation. In addition, equity issues play a major role in risk perception. The more risks are seen as unfair for the exposed population, the more they are judged as severe and unacceptable (Kasperson and Kasperson 1983; Short 1984). The most influential factors are listed in Table 6.

Table 6. **LIST OF IMPORTANT QUALITATIVE RISK CHARACTERISTICS**

Qualitative Characteristics	Direction of Influence
• Personal control	increases risk tolerance
• Institutional control	depends on confidence in respective institution
• Voluntariness	increases risk tolerance
• Familiarity	increases risk tolerance
• Dread	decreases risk tolerance
• Inequitable distribution of risks and benefits	depends on individual utility, strong social incentive for rejecting risks
• Artificiality of risk source	amplifies attention to risk, often decreases risk tolerance
• Blame	increases quest for social and political responses

- *the beliefs associated with the cause of risk:* The perception of risk is often part of an attitude that a person holds about the cause of the risk, i.e. a technology, human activity, or natural event. Attitudes encompass a series of beliefs about the nature, consequences, history, and justifiability of a risk cause (Thomas et al. 1980; Otway and Thomas 1982). Due to the tendency to avoid cognitive dissonance, i.e. emotional stress caused by conflicting beliefs (Festinger 1957), most people are inclined to perceive risks as more serious and threatening if the other beliefs contain negative connotations and vice versa. Often risk perception is a product of these underlying beliefs rather than the cause for these beliefs (Clarke 1989).

It should be noted that estimations of seriousness (risk magnitude) and judgments about acceptability are closely related in risk perception. Most people integrate information about the magnitude of the risk and qualitative factors (e.g., the fairness of the risk situation) into their overall judgment about the (perceived) seriousness of the respective risk.

4.4 Semantic images

Another approach within qualitative risk perception studies is looking at "semantic images". With respect to the meaning of technological risk, four distinct semantic images have been described in Renn 1990a. These concepts are summarized in Table 7. In addition to these models, additional images of risk can be identified for natural disasters and lifestyle risks.

- *Risk as a pending danger (Damocles sword):* Risks are seen as random threats that can trigger a disaster without prior notice and without sufficient time to cope with the hazard involved. This image is linked to artificial risk sources with large catastrophic potential.

Table 7. **THE FOUR SEMANTIC IMAGES OF RISK IN PUBLIC PERCEPTION**

1. Pending Danger (Damocles Sword) • artificial risk source • large catastrophic potential • inequitable risk-benefit distribution • perception of randomness as a threat
2. Slow Killers (Pandora's Box) • (artificial) ingredient in food, water, or air • delayed effects; non-catastrophic • contingent on information rather than experience • quest for deterministic risk management • strong incentive for blame
3. Cost-benefit Ratio (Athena's Scale) • confined to monetary gains and losses • orientation towards variance of distribution rather than expected value • asymmetry between risks and gains • dominance of probabilistic thinking
4. Avocational Thrill (Hercules Image) • personal control over degree of risk • personal skills necessary to master danger • voluntary activity • non-catastrophic consequences

- *Slow killers (Pandora's box):* Risk is seen as an invisible threat to one's health or well-being. Effects are usually delayed and affect only few people at the same time. Knowledge about these risks is based on information by others rather than on personal experience. These risks pose a major demand for trustworthiness in those institutions that provide information and manage the hazard.
- *Cost-benefit ratio (Athena's scale):* Risks are perceived as a balancing of gains and losses. This concept of risk comes closest to the technical understanding of risk. However, this image is only used in peoples' perceptions of monetary gains and losses.
- *Avocational thrill (Hercules image):* Often risks are actively explored and desired (Machlis and Rosa 1990). These risks include all activities for which personal skills are necessary to master the dangerous situation. The thrill is derived from the enjoyment of having control over one's environment or oneself. Such risks are always voluntary and allow personal control over the degree of riskiness.

Such semantic images allow individuals to order risks on the basis of a few salient characteristics. Reducing complexity by creating classes of similar phenomena is certainly a major strategy for coping with information overload and uncertainty. Thus semantic images are used as 'guides' to navigate through an abundance of often contradicting information and to balance the time for collecting and processing information with the personal need for orientation and attitude formation. However, they don't reflect the complexity of hazard sources and outcomes - and therefore might burden the communication about controversial risks.

4.5 Differences between hazard types

As hazards differ in terms of sources and impacts, principal disparities are to be expected for both perceived risk level and acceptance of risks.

- *Risk magnitude ratings:* Particular risk sources are more or less always evaluated as high risks. According to a summary of 13 studies from 11 different countries (cf. Rohrmann 1999, Box G), for most people the worst risks are smoking and nuclear power, followed by asbestos production and automobiles. Alcohol/spirits and various chemical products such as pesticides also get high risk magnitude judgments. Interestingly, the two risks rated highest include the best researched one, smoking impacts, and the most debated one, the use of nuclear energy, and both have very high publicity.
- *Acceptance of different risk types:* How people evaluate risk sources, and to what extent they are prepared to accept a risk, are dependent on the type of risk and the reasons they are exposed to that hazard. In particular two factors have been discussed in this context: whether the risk is taken voluntarily or not (an issue first raised by Starr 1969), and whether the source is "natural" versus human-made or technology-induced (see, e.g., Brun 1992). In his review of psychometric risk perception studies, Rohrmann (1999) compared results for five types of hazards and concluded that "lifestyle" hazards tend to get the highest risk magnitude ratings, followed by technology-induced hazards; the lowest ranks are for natural hazards and occupational risk exposure; ratings for sport-related risks vary widely.

There are also obvious differences in acceptance judgments: Individual risk acceptance tends to be higher for private, societal risk acceptance higher for occupational hazards. Regarding residential environmental risks, risk acceptance is higher for natural hazards in both a societal and an individual perspective. In this context it should be noted that many hazards presented in risk perception studies are hard to classify, for example: is "cancer" natural, a lifestyle outcome or perhaps the result of technology usage? Are "floods" natural or a human-made disaster (because of settlements or dam construction); is living in such an area a voluntary risk?

In sum, the characteristics of a risk source clearly matter when people judge and assess risks. Yet, the conclusion that "voluntary" or "natural" risks are more acceptable is too simple, as these criteria are confounded by other aspects, such as local/global, private/occupational, present/future impacts, and so on. Within individual activities, acceptance appears to be mostly motivated by benefit considerations and control beliefs. Within risks resulting from environmental influences, hazards caused by "nature" indeed seem to be perceived and evaluated as more tolerable than those stemming from human endeavors or technologies, even though neither objective risk assessments nor subjective risk magnitude ratings might differ that much between the two kinds of risks.

4.6 The structure of risk aspects

The idea that few general concepts determine how humans assess hazards has been pursued systematically in psychometric research. There are basically two types of studies, ones that look at dimensions and patterns of risks (using multi-dimensional scaling or factor analysis or clustering) and those that utilize regression techniques and structural/causal modelling. A main finding similarly presented in many studies is the factorial structure of risk concepts. Usually two or three factors are found, representing:

<1> the dreadfulness of risks,

<2> the degree of knowledge of and familiarity with the hazard,

<3> (in some studies) the number of exposed people.

Fig. 3 shows a 2-dimensional structure frequently published by Slovic, Fischhoff & Lichtenstein (e.g. 1980, 1981, 1985, 1986) and Slovic (1987, 1992). Several studies repeating the approach by Slovic et al. produced a roughly equivalent structure, at least for factors <1> and <2>, (e.g., Englander et al. 1986, Goszczynska et al. 1991, Jianguang 1994, Kleinhesselink & Rosa 1991, Teigen et al. 1988). Other researchers emphasize different and/or additional dimensions, e.g., Brun 1992 ("novelty" factor), Mullet et al. 1993 ("evaluative" factor), Puy & Aragones 1994 ("chronicity").

However, it should be remembered that factor-analytic solutions can be 'tailored' to some degree. Also, changing the set of risks and in particular the method of identifying dimensions (cf., e.g., the studies by Bastide et al. 1989, Hohenemser et al. 1983, Johnson & Tversky 1984, Puy & Aragones 1994, Vlek & Stallen 1981) yields deviant results. Actually the graphs usually presented showing a set of risk sources in a two-dimensional space may result either directly from correlations (or similarity data) between sources or (indirectly) from factor analyses of correlations between risk aspects, from which then factor scores for sources can be derived.

Figure 3. **TWO-DIMENSIONAL FACTOR SPACE FOR HAZARDS**

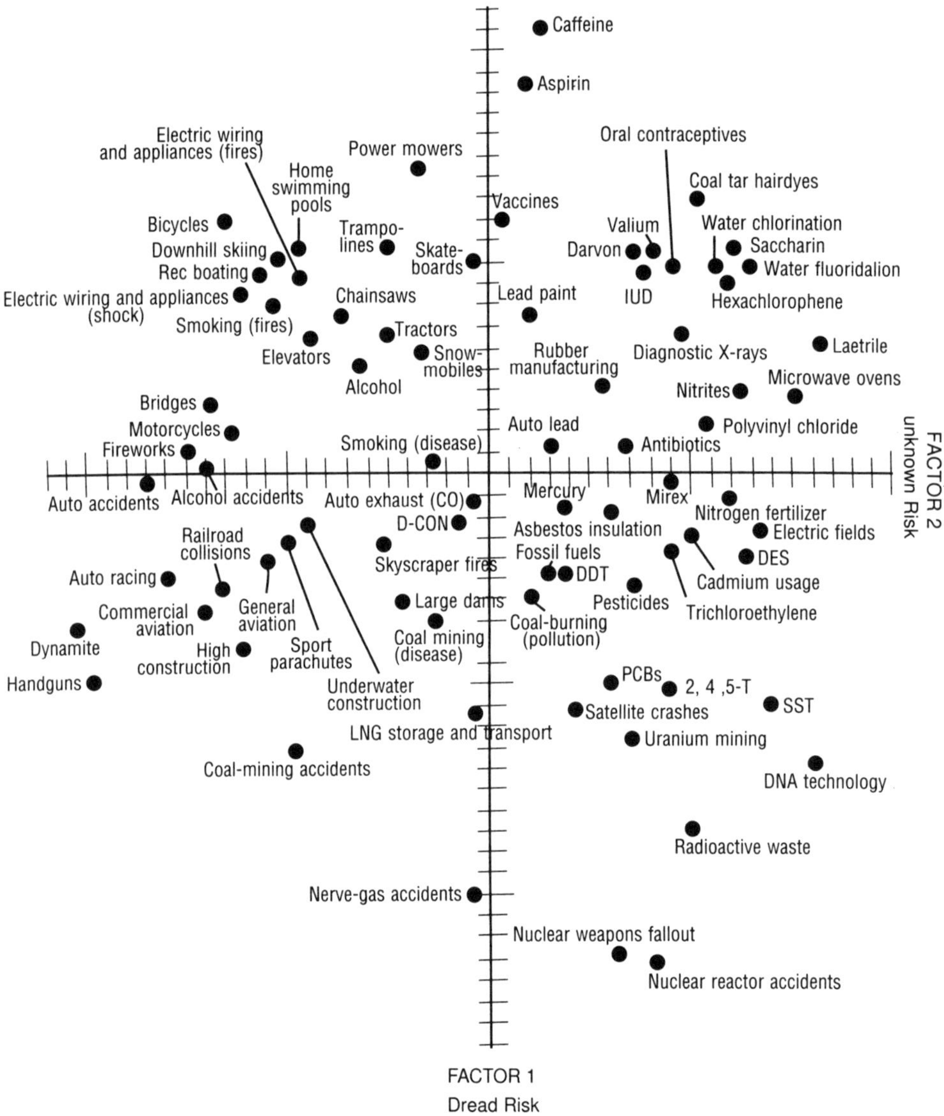

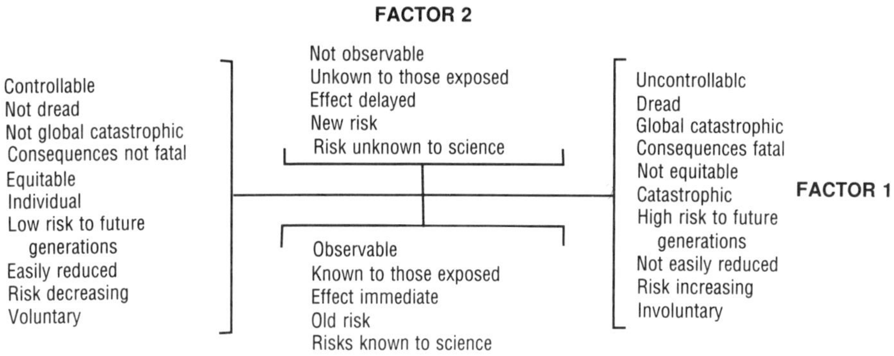

Location of eighty-one hazard on Factor 1 and 2 derived from the relationship among eighteen risk characteristics. Each factor is made up of a combination of characteristics, as indicated by the lower diagram.
(Slovic, Fischhoff, and Lichtenstein 1985)

5 CORE FINDINGS ON THE FACTORS INFLUENCING RISK EVALUATIONS

The following sections describe factors that shape the direction and the intensity of risk evaluations. They can be regarded as independent variables in the complex perception process. Only a selection of factors can be dealt with; the focus will be on trust, social values, cultural and political structure, and amplification processes.

5.1 Institutional trust and confidence

Among the most influential social factors in shaping risk perception and responses social networks and reference group judgments are particularly influential since most information about risk is not learned through personal experience but through "second-hand" learning. With the advent of ever more complex technologies and the progression of scientific methods to detect even the smallest quantities of harmful substances, personal experience of risk has been more and more replaced by information about risks and individual control over risk by institutional risk management. As a consequence, people rely more than ever on the credibility and sincerity of those from whom they receive information about risk. Thus, trust in institutional performance has been a major key for risk responses.

Trust on a personal level is a subjective exception that a person will refrain from behavioral options that may harm the trusting person. Trust necessarily entails risk-taking, but, in contrast to the scientific endeavor of predicting the probability of potential outcomes, trust implies that the selection of options is left to the entrusted person or institution (Luhmann 1980, 1973). This saves time and effort. On a more aggregate level, trust denotes a generalized medium of social differentiation and division of labor (Parsons 1960). The performance of specialized institutions in economy and government relies on a prior investment of trust by those who are served by this institution or finance its functioning (Durkheim 1933; Luhmann 1973).

It is obvious that modern societies face difficulties in providing sufficient trust for reaching consensus on its complex and differentiated activities. All public institutions have lost trust and credibility over the last two decades except for the news media (Lipset and Schneider 1983; Peters 1991, 1999).

Trust and credibility losses are high for industry, the political system, and many government agencies. Science still has a high degree of credibility although much less than two decades ago. Most sociologists believe that the decline of confidence in public institutions is partially a function of better education and the increase of public aspirations with respect to their share of public resources and welfare (Renn and Levine 1991). In addition, the complexity of social issues and the pluralization of values and lifestyles may have contributed to a growing dissatisfaction with the actual performance of institutions (Renn 1995).

Lack of trust does not indicate, however, a declining relevance of trust for governing modern societies and managing technological risks. The contrary is true. The reliance of the technological society on trustful relationships between and among its subsystems has never been stronger than today. However, such a need for trust makes people more and more sensitive towards situations in which their investment of trust has been factually or allegedly misguided. The more trust is needed for implementing cooperative efforts or for coping with external effects of social actions, the more cautious people are in assigning credibility to those whom they are supposed to trust.

This is particularly relevant for risk issues. Since the notion of risk implies that random events may trigger accidents or losses, risk management institutions are always forced to legitimate their action or inaction when faced with an accident. On one hand they can cover up mismanagement by referring to the alleged randomness of the event (labeling it as unpredictable outcome or an act of God), on the other hand they may be blamed for events for which they could not possibly provide protective actions in advance (Luhmann 1986). The stochastic nature of risk demands trustful relationships between risk managers and risk bearers, since single events do not prove nor disprove management failures; at the same time they provoke suspicion and doubt.

In risk debates issues of trust evolve around institutions and their representatives. People's responses to risk depend among others on their confidence that they have in risk initiating and controlling institutions. The handling of risk by private corporations and governmental agencies has been crucial for explaining the mobilization rate of individuals for taking actions. The more individuals believe that risks are not properly handled (in addition to being perceived as serious threats) the higher is the likelihood that people will be politically active. Negative attitudes are one necessary but by far not a sufficient reason for behavioral responses. Public confidence in institutional performance is another and even more important element in triggering behavioral responses.

5.2 Value commitments

Positions towards risks of technologies or human activities differ as a result of divergent views about the goal(s) and values that are to be accomplished by providing risk-related technologies for production, consumption or distribution. In the social sciences, values are placed in clusters that seem to belong together although most people are characterized by mixed values systems (Fiorino 1989). These clusters are summarized in Table 8.

Table 8. **VALUE CLUSTERS AND THEIR SOCIAL AND CULTURAL FUNCTIONS**

Cluster Name	Examples	Function
traditional values	patriotism, regional or ethnic identity, social status, family stability	group and cultural identity
work ethics	diligence, punctuality, efficiency, discipline, deferred gratification	functionality, efficiency
hedonistic values	consumption, enjoyment, fun, immediate gratification	incentive, motivation
postmaterialistic values	harmony, social responsibility, environmental quality, decentralization, quality of life	moral legitimation, cultural commitment

In contrast to many popular views, there is not a universal shift towards post-materialistic values throughout the western world (Klages 1984). It is true that these values have become more important and can be found on the value priority list of almost every individual, but the claim of a new post-materialistic personality is total fiction. Most people demonstrate a mix of all value clusters depending on context and social relations. A vast majority of people is still interested in gaining additional personal income (even if they rate it low on the scale of personal aspirations). Even unfashionable virtues such as discipline and efficiency have their place in most people's value portfolio. However, many of the traditional and work-related values are withdrawn from situations in which they used to be the dominating orientations. This has been the case with many technologies: they were perceived as manifestations of work ethics and hedonistic values (production and consumption), but are increasingly related to post-materialistic concerns. This shift in value application is partially responsible for the perception of ambivalence, which is so typical for modern attitudes towards technologies.

Since values have an impact on technology evaluation, responses to risk rely on the requirement that they load high on each value cluster except for the traditional values. Traditional values are normally disassociated with the use of technologies. Positive responses to risk-related technologies or activities depend on the possibility of the evaluator to forge links between the three value clusters and the development or use of the technologies under consideration. That is why technology design and development should acknowledge and incorporate the likely implications of new products or processes for the three major value clusters. If the perceived implications meet all three clusters, acceptance problems are unlikely to play a major role in the diffusion process.

5.3 Cultural group affiliations

The first wave of risk perception research was - at least implicitly - based on the assumption that risk judgements and evaluations are "universal" processes, independent of, e.g., social status, professional affiliation, ideologies, national heritage and other cultural contexts. This view can easily be questioned though, based on both conceptual and empirical grounds.

Social psychologists as well as sociologists and anthropologists have theorized that the evaluation and acceptance of risks is influenced by an individual's affiliation to relevant groups within a culture - the "worldview" of such (sub-)cultures is assumed to shape the response to risk situations. This perspective has been pursued in two - rather separate - research streams, one based on a 'cultural' typology, one on professional and political group affiliations.

- *Cultural approaches based on Grid/Group typology:* Anthropologists and cultural sociologists such as Douglas and Wildavsky 1982, Rayner and Cantor (1987) or Thompson, Ellis and Wildavsky (1990) have postulated four or five patterns of value clusters that separate different cultural groups from each other. These different groups are assumed to have formed specific positions on risk topics and have developed corresponding attitudes and strategies. They differ in the degree of group cohesiveness (the extent to which someone finds identity in a social group), and the degree of grid (the extent to which someone accepts and respects a formal system of hierarchy and procedural rules). The main prototypes have been described as follows: There are four major groups in modern society that are likely to enter the risk arena: the entrepreneurs, the egalitarians, the bureaucrats, and the stratified individuals. Some authors have included a fifth group, the autonomous individuals who are literate in several risk cultures (e.g., Schwarz & Thompson 1990). All five groups can be localized within the group-grid space (see Figure 4). Organizations or social groups belonging to the entrepreneurial prototype perceive risk taking as an opportunity to succeed in a competitive market and to pursue their personal goals (Rayner 1987). They are characterized by a low degree of hierarchy and a low degree of cohesion. This group contrasts most with organizations or groups belonging to the egalitarian prototype, which emphasizes cooperation and equality rather than competition and freedom. Egalitarians are also characterized by low hierarchy, but have developed a strong sense of group cohesiveness and solidarity. When facing risks they tend to focus on long term effects of human activities and are more likely to abandon an activity (even if they perceive it as beneficial to them) than to take chances.

The third prototype, i.e. the bureaucrats, relies on rules and procedures to cope with uncertainty. Bureaucrats are both, hierarchical and cohesive in their group relations. As long as risks are managed by a capable institution and coping strategies have been provided for all eventualities, there is no need to worry about risks. The fourth prototype, the group of atomized or stratified individuals, principally believes in hierarchy, but does not identify with the hierarchy to which they belong. These people trust only themselves, are often confused about risk issues, and are likely to take high risks for themselves, but oppose any risk that they feel is imposed on them (Thompson 1980). A final group is the group of the autonomous individuals in the center of the group-grid coordinates. Thompson describes autonomous individuals as self-centered hermits and short-term risk evaluators. They could be seen as potential mediators in risk conflicts, since they build multiple alliances to the four other groups and believe in hierarchy only if they can relate the authority to superior performance or knowledge (Renn 1992b).

Figure 4. **THE FIVE PROTOTYPES IN CULTURAL THEORY**

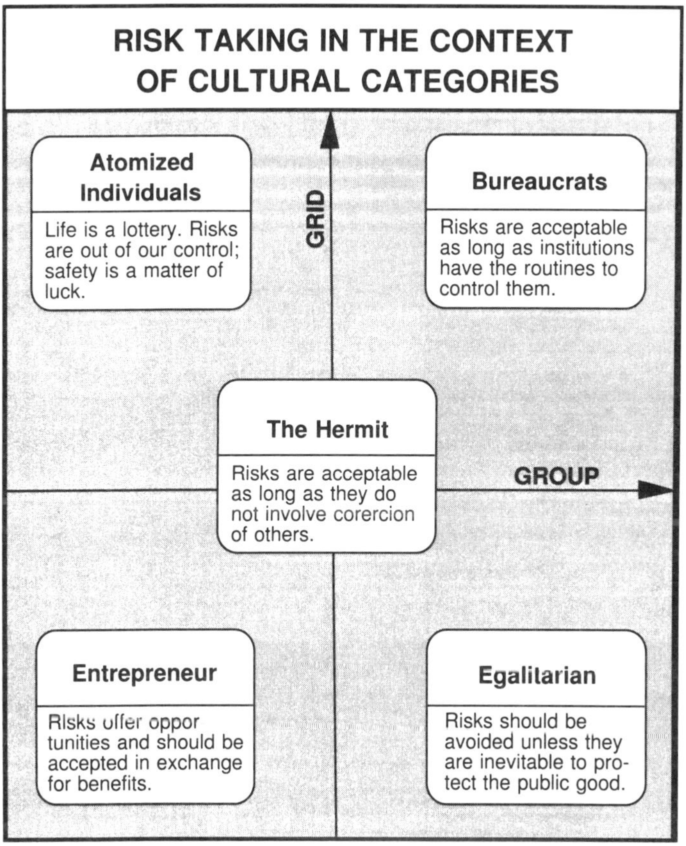

This "cultural" theory has been criticized on several grounds (Nelkin 1982, Johnson 1987, Renn 1992a, Sjöberg 1996,1997). First, most authors within the cultural theory emphasize that cultural prototypes do not characterize individuals but social aggregates. Second, the relationship between cultural prototype and organizational interest is unclear and problematic. If cultural affiliation precedes interest, then what determines to which cultural prototype groups or organizations belong? Third, the selection of the five prototypes as the only relevant cultural patterns in modern society needs more evidence than the reference to tribal organizations (Douglas 1985) or generic models of human interactions (Wildavsky and Dake 1990). Furthermore, if prototypes are mixed in organizations, then the perspective (similar to many sociological concepts) cannot be falsified. Any observed behavior is compatible with some mix of prototypes. Lastly, the cultural perspective has not provided sufficient empirical evidence of its validity. For example, Dake (1991) compared the risk perception for three "cultural biases", namely individualism, egalitarianism, and hierarchy. It turned out that "egalitarians" are more risk-aversive than other political cultures and reveal more concern about technological and environmental hazards; effects were not very strong though. A thorough examination of cultural risk theory by Marris et al. 1996 found only a small influence of four worldviews (individualism, fatalism, hierarchy, egalitarianism) on risk perception. Results presented by Peters & Slovic 1996 and Sjöberg 1997 are not favorable either.

The cultural theory of risk has its shortcomings and its merits. The reduction of cultural clusters to basically three important prototypes (entrepreneurial, egalitarian, and bureaucratic) may be a cogent and intuitively plausible proposition in analyzing risk responses, but it should be treated as a hypothesis rather than the exclusive explanation. The emphasis on values and worldviews rather than interests and utilities (which in themselves are reflections of one's worldview) can be seen as a major accomplishment of this theory.

- *Professional and political/ideological affiliation:* A couple of researchers have looked at cross-cultural differences in risk perception based on empirical group features and identified several significant societal factors. Some examples shall be briefly mentioned. In a small exploratory study, Belrose & Pilisuk (1991) compared three occupational groups: firefighters, insurance salespeople, and radiation specialists. They found differences in risk judgments, linked to the respondents' occupation ("vocation does appear to provide a framework for risk perception").

In a series of studies conducted in four countries (Australia, New Zealand, Germany, China; see Borcherding et al. 1986, Rohrmann 1994, Rohrmann, this volume), four specific groups, defined by a combination of profession and political/societal positions, were looked at, yielding considerable differences in risk perception:

- People with an "ecological orientation" as well as those involved in "feminist" issues evaluate risks most critically on all riskiness scales while respondents with a "technological orientation" (e.g., engineers) show the lowest risk ratings, see more benefits and are more ready to accept risks. The judgments of a "monetarian" group are in between the extremes.
- This pattern is most obvious for technology-induced risks, and the results are quite similar across the three countries included in this research.
- The subgroups of professionals and students in each of the pertinent fields come up with quite similar risk evaluations.

Effects of occupational factors have been found by others as well (e.g., Gould et al. 1988, Nyland 1993, Sjoberg et al., this volume, Tiemann & Tiemann 1985, Vlek & Stallen 1981). Clearly professional and political affiliations are a major facet of people's 'worldviews' and thus relevant for the subjective interpretation of risks.

5.4 Cross-national differences

There are about twenty cross-cultural risk perception studies based on cross-national data collection (cf. Table 4 above); a few studies also looked at cultural differences within countries by comparing societal groups (cf. section 4.7). Documentation for both types is provided in Rohrmann 1999.

All cross-national studies find differences in perceived risk judgments. This mainly applies to mean risk magnitude ratings for specific risk sources which are evaluated quite disparately, depending on the political, social and economic context of each nation. For example, the perception of nuclear-energy risks as well as smoking differs significantly in countries such as France, Sweden, Germany, China, Japan, USA, Brazil etc. An example is the study by Hinman et al. (1993), a Japan/USA comparison (cf. Table 9).

Table 9. **A JAPAN/USA COMPARISON OF RISK PERCEPTION (Hinman et al. 1993)**

Respondents:	Tokyo/Japan: N=290; Pacific Northwest of the USA: N=747; Time of data collection: 1988, 1990
Hazards:	30 activities, substances and technologies, including smoking (S), asbestos (A), accidents involving alcohol (C), DNA technology (D), nuclear energy (N), nuclear war (W).
Risk aspects:	four criteria, including „feelings of dread“
Mean ratings:	Mostly higher for japanese than US-american data Japan sample SC D AN W *Response scale:* 1—2—3—4—5—6—7 USA sample D NA CS W
Comments:	Large differences in health hazards (e.g. smoking) and tho risk of war; results for technologies mostly equivalent
Sample validity:	Population samples but much lower response rate in Japan (29%) than the U.S. (53%).

However, the correlational structure of risk aspects found in the various studies is altogether similar; also, an homologous factor structure of hazards has been identified in several countries.

Findings on cross-cultural differences have to be assessed carefully, as most samples were not representative of the population under investigation. Nevertheless the results indicate considerable cultural influences on individual and societal risk evaluation and certainly question the appropriateness of generalizing the various 'national' studies. Consequently, single risk perception studies, bound to a specific cultural context, have restricted external validity.

Finally, it seems that cross-national differences in risk perception are less stringent than societal group disparities (as presumed in Rohrmann 1994). However, more evidence is needed to validate this interpretation.

5.5 Social Amplification of Risk

In 1988, Kasperson and colleagues proposed a novel approach to study the social response to risk. The concept of social amplification of risk is based on the thesis that events pertaining to hazards interact with psychological, social, institutional, and cultural processes in ways that can heighten or attenuate individual and social perceptions of risk and shape risk behavior. Behavioral patterns, in turn, generate secondary social or economic consequences that extend far beyond direct harms to human health or the environment, including significant indirect impacts such as liability, insurance costs, loss of confidence in institutions, or alienation from community affairs (Kasperson et al. 1988).

Such secondary effects often trigger demands for additional institutional responses and protective actions, or, conversely (in the case of risk attenuation), place impediments in the path of needed protective actions. In accordance with the metaphor of amplification in the processing of electronic signals, amplification includes both intensifying and attenuating signals about risk. Thus, alleged overreactions of target audiences receive the same attention as alleged 'downplaying'.

Figure 5 illustrates the process of amplification. The amplification process starts with either a physical event (such as an accident) or the recognition of an adverse effect (such as the discovery of the ozone hole). In both cases, individuals or groups will select specific characteristics of these events or aspects of the studies and interpret them according to their perceptions and mental schemes. These interpretations are formed into a message and communicated to other individuals and groups (Renn 1991). Individuals or groups collect and respond to information about risks and act in our terminology as "amplification stations" through behavioral responses or communication. Amplification stations can be individuals, groups, or institutions. Amplification differs among individuals in their roles as private citizens and in their roles as employees or members of social groups and public institutions.

The behavioral and communicative responses are likely to evoke secondary effects that extend beyond the people directly affected by the original hazard event. Secondary impacts are, in turn, perceived by social groups and individuals so that another stage of amplification may occur to produce third-order impacts. The impacts may spread or "ripple" to other parties, distant locations, or other risk arenas. Each order of impact will not only disseminate social and political impacts but may also trigger (in risk amplification) or hinder (in risk attenuation) positive changes for risk reduction.

Drawing upon this concept of social amplification of risk, Renn et al. (1992) investigated the functional relationships among five sets of variables that enter into the amplification process. The first class of variables included the physical consequences of 128 hazardous events (events that exposed humans or the environment to physical harm); the second class referred to the amount of press coverage about these 128 events; the third class entailed the individual layperson perceptions with respect to these events; the fourth class described the public responses (individual behavioral intentions and group mobilization potential) to these hazards; and the fifth class contained the socioeconomic and political impacts of these events as measured by documents and a Group Delphi with experts. The study investigated the structure of causal relationships among these variables classes.

Figure 5. **THE CONCEPT OF SOCIAL AMPLIFICATION OF RISK**

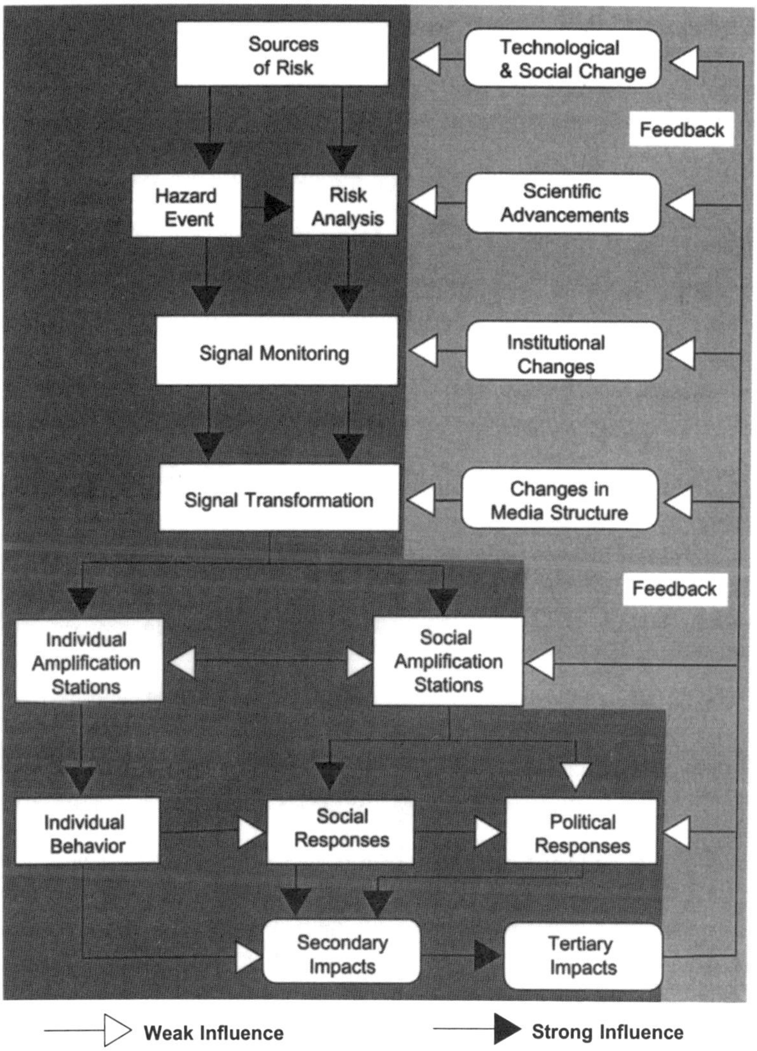

The most interesting result of this study is the weak link between casualties and most of the other variables. The best physical risk predictor is exposure rather than any other indicator of harm. Exposure contributes to dread and is also highly correlated with media coverage. Its direct influence on intended action is small indicating that exposure operates through risk perception variables to influence personal actions. The link between exposure and societal impacts is not significant, however, despite the initially high correlation between the two variables (Renn et al. 1992). Exposure appears to shape societal experiences with risk through the media and through perceptions and intended individual actions. The data reflects the major assumption of the social amplification model. i.e., that physical events are observed and interpreted by groups and individuals, amplified through individual and social processors, and then expressed in terms of societal consequences.

The social amplification framework provides an integrative concept. The distinction between individual and social amplification stations corresponds with the two traditions in risk perception: the individual processing of information and the social responses to risk based on experience of (dis)trust, social values, and cultural affiliations. It provides a more holistic picture of the risk perception process and takes into account psychological, sociological, and cultural aspects.

5.6 Determinants of risk evaluations - synopsis

The determinants of risk judgments and acceptance ratings are plenty and interrelated. The results gained from correlational analyses and multiple regressions demonstrate that socio-psychological factors in particular have substantial influence on the evaluation of risks (e.g., reasons for exposure, perceived controllability, attitudes such as environmental concern or technology skepticism, professional orientation, fear associations, beneficial potential in individual or societal terms). On the other hand, 'technical' criteria (e.g., fatality probabilities, safety characteristics) are less relevant. The subjective concept of "risk" should be seen as a multifaceted/ multidimensional construct. While the majority of risk perception studies focus on negative impacts, the complex relationship between perceived riskiness and benefits of risk sources has gained increased interest (e.g., Alhakami & Slovic 1994, Borcherding et al. 1986, Gardner & Gould 1989, Gregory & Mendelsohn 1993, Holtgrave & Weber 1993, Rohrmann 1991c). However, the subjective trade-off between quantitative and qualitative risk and benefit features (see Gregory & Lichtenstein 1994) seems not yet sufficiently understood.

In Figure 6 the respective findings - particularly those from structural/causal modeling - are 'condensed' into an influence diagram (adapted from Rohrmann 1999). The structure shown in this model is gained from analyses with a variety of risk sources and general samples. For specific risks and social groups the result could be quite different. For example, a variable such as 'catastrophic potential' seems to be relevant mainly for risks beyond individual control while 'probability of dying' might be considered for personal voluntary activities; or general value orientations particularly influence judgments of technology-induced risks. Indeed, each individual may have a personal influence pattern for the relevance of variables in this model (as can be shown in regression analyses for individual data, e.g., Borcherding et al. 1986). This begins

Figure 6. **SUBJECTIVE EVALUATION OF RISK: STRUCTURAL MODEL**

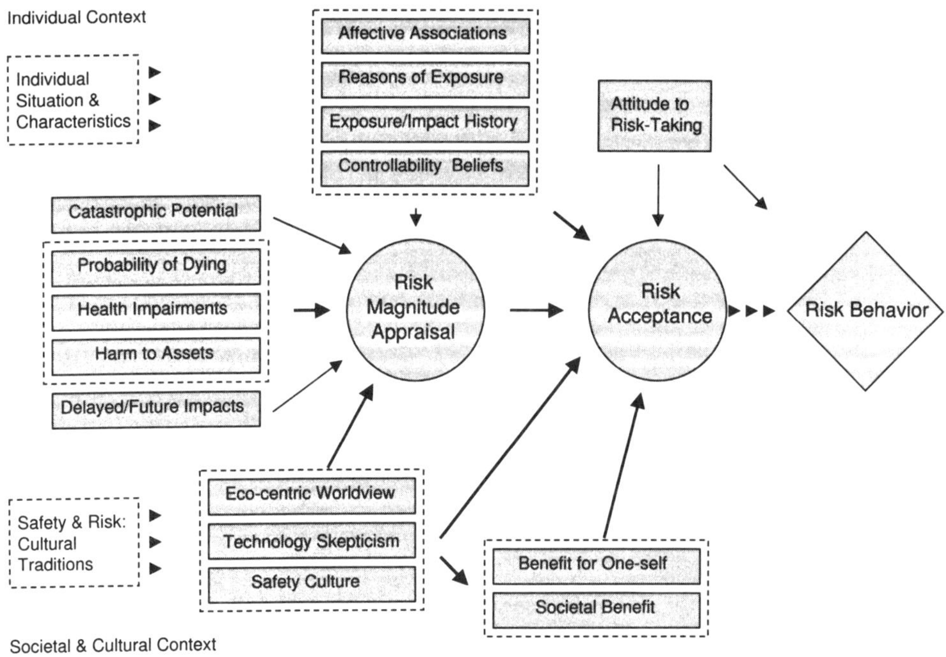

with the intuitive risk definition a person employs (Drottz-Sjöberg 1991) and ends with the importance of societal attitudes not specific to the risk source (see e.g., Dake 1991, Rohrmann 1994).

Altogether such structural analyses reveal that risk perception - both the rating and the evaluation of hazards - is influenced by a multitude of individual and societal factors. Neither the 'classic' hazard attributes (i.e., probability and adversity of negative outcomes) nor a typology of 'worldviews' (e.g., societal prototypes) is sufficient in itself to explain how people judge risks.

The richness and complexity of individual and social factors that shape risk perception demonstrate that the intuitive understanding of risk is a multidimensional concept and cannot be reduced to the product of probabilities and consequences (Allen 1987). Although risk perceptions differ considerably among social and cultural groups, the multi-dimensionality of risk and the integration of beliefs related to risk, the cause of risk, and its circumstances into a consistent belief system appear to be common characteristics of public risk perception in almost all countries in which such studies have been performed. Furthermore, the experience of risk is not limited to the threat of facing harm in the future. It includes subjective predictions of possible outcomes, the social and cultural context in which the risk is experienced, the mental images the risk situation evokes, the perception of the players who are involved in the risk situation and the judgments about fairness and equity related to the distribution of potential hazardous events. In this sense, risk is a social construct rather than a physical entity. However, this construct is not without physical foundation: people do face and fear disaster and natural degradation. Experiencing or expecting these physical changes are part of the social experience of risk, but it is not the only, in some instances not even the most important part of the social perception of risk.

6 RELEVANCE OF RISK PERCEPTION STUDIES

In conclusion: Obviously 'technical' and solely 'quantitative' approaches (as prevalent in risk analysis) are inadequate to reflect the complex pattern of individual risk perception. (This has been stressed by many authors, see, e.g., Binswanger 1990, Covello 1983, Fischhoff 1989, Fischhoff et al. 1984, Freudenburg 1988, Heimer 1988, Jungermann & Slovic 1993, O'Riordan 1983, Rayner 1987, Renn 1983, 1993, Rohrmann 1995, Sjöberg 1987, Slovic 1991, Slovic et al. 1982, Wildavsky 1988, v. Winterfeldt & Edwards 1984). How people think about the magnitude and the acceptability of risks and how they make their respective judgments and decisions is influenced by knowledge, values and feelings. They include a variety of 'qualitative' consequences of risky activities or living conditions beyond hazard scenarios, probabilistic information about accidents, mortality rates and so on. The mental models and other psychological mechanisms which people use (e.g., cognitive heuristics; (dis-)trust in information sources; risk images) are internalized through social and cultural learning and constantly moderated by media reports, peer influences and other amplification processes.

Laypersons' views on risks obviously are intuitive and less formal and precise than experts' statements, "however, their basic conceptualization of risk is much richer than that of experts and reflects legitimate concerns that are typically omitted from expert risk assessments" (Slovic 1987). In fact, risk judgments indicate more than just perception of riskiness - they reveal global views on the development of humankind, on technological progress, on the meaning of nature, and on the 'fair' distribution of chances, benefits and risks.

What are the implications for handling risk problems by individuals, institutions and governmental authorities? The understanding of risk perception is relevant to hazard prevention, risk management and safety enhancement in several ways. The essential link is risk communication, i.e., the exchange of information about risk issues between involved parties. On the one hand this - rather new and still growing - field of risk research is considerably influenced by findings from social-scientific risk perception research. On the other hand knowledge about the subjective experience and evaluation of risks is indispensable for analyzing, aiding and improving risk communication (Covello, McCallum & Pavlova 1989, Femers 1993, Fischhoff 1994, Haller 1990, Jungermann, Kasperson & Wiedemann 1988, Kasperson & Stallen 1990, Leiss 1990, Plough & Krimsky 1987, Renn 1990b, Rohrmann 1992).

Whether the particular aim of risk communication is (A) advancing knowledge, (B) behavior modification or (C) co-operative conflict resolution, the awareness and consideration of people's existing risk perception is a precondition for successful work. This is particularly true in dealing with societal controversies about the evaluation and management of risks. Conflict analysis and resolution must be based on mutual knowledge and acceptance of the actor's way of thinking about risk (Renn 1992b, 1992c, Rohrmann 1991a, Vlek & Cvetkovich 1989, Wiedemann et al. 1991).

Finally, social-scientific expertise about risk perception can be utilized for many purposes and even misused, e.g., for 'selling' risky enterprises or 'shifting' attitudes and values, for calming people down or discrediting opponents, and so on (for critical views see Earle & Cvetkovich 1988; Lopes 1992; Stern & Fineberg 1996). Nevertheless, such expertise is indispensable if efforts to improve individual responses to risks or to facilitate societal decision making about risk policies in a 'rational discourse' are to be based on people's actual knowledge, feelings and appraisals of the respective risks. However, the understanding of risk perception and its relation to other aspects of human behavior and social processes is not yet sufficiently coherent. Ongoing research is needed to advance the theoretical foundation of the field as well as the practical utilization of findings.

7 REFERENCES

Alhakami, A. S. & Slovic, P. 1994. A psychological study of the inverse relationship between perceived risk and perceived benefit. Risk Analysis, 14, 1085-1096.

Allen, F. W. 1987. Towards a holistic appreciation of risk: the challenge for communicators and policymakers. Science, Technology, and Human Values, 12, 138-143.

Appelbaum, R. P. 1977. The future is made, not predicted: technocratic planners vs. public interests. Society, 5, 49-53.

Arabie, P. & Maschmeyer, C. 1988. Some current models for the perception and judgment of risk. Organizational Behavior and Human Decision Processes, 41, 300-329.

Bastide, S., Moatti, J.-P., Pages, J.-P. & Fagnani, F. 1989. Risk perception and social acceptability of technologies: the french case. Risk Analysis, 9, 215-225.

Bechmann, G. (Hg.) 1993. Risiko und Gesellschaft. Opladen: Westdeutscher Verlag.

Beck, U. 1992. Risk Society: Towards a new modernity. London: Sage. <Original: Beck, U. 1986. Die Risikogesellschaft: auf dem Weg in eine andere Moderne. Frankfurt/Main: Suhrkamp.>

Bellrose, C. & Pilisuk, M. 1991. Vocational risk tolerance and perceptions of occupational hazards. Basic and Applied Social Psychology, 12, 303-323.

Binswanger, H. C. 1990. Neue Dimensionen des Risikos. Zeitschrift für Umweltpolitik und Umweltrecht, 13, 103-118.

Borcherding, K., Rohrmann, B. & Eppel, T. 1986. A psychological study on the cognitive structure of risk evaluations. In: Brehmer, B., Jungermann, H., Lourens, P. & Sevon, G. (Eds.): New directions in research on decision making, Amsterdam: North- Holland, 245-262.

Bostrom, A. 1990. A mental models approach to exploring perceptions of hazardous processes. Diss. Carnegie Mellon University.

Bostrom, A., Fischhoff, B. & Morgan, M. G. 1992. Characterizing mental models of hazardous processes: A methodology and an application to radon. Journal of Social Issues, 48, 85-100.

Bradbury, J. A. 1989. The policy implications of differing concepts of risk. Science, Technology, and Human Values , 14 , 380-399.

Brehmer, B. 1987. The psychology of risk. In: Singleton, W. T. & Hovden, J. (Eds.): Risk and decision, New York: Wiley, 25-39.

Brehmer, B. 1994. Some notes on psychological research related to risk. In Brehmer, B. & Sahlin, N. E. (Eds.): Future risks and risk management, pp. 79-91. Amsterdam: Kluwer.

Brengelmann, J. C. 1991. Die Lust auf Spiel und Risiko. Zürich: Varia Press.

Brown, J. M. 1988. Psychological aspects of environmental risks from industrial and nuclear power plants. In: Sime, J. D. (Ed.): Safety in built environment, London, New York: E. & F.N. Spon, 337-347.

Brun, W. 1992. Cognitive components in risk perception: Natural versus manmade risks. Journal of Behavioral Decision Making, 5, 117-132.

Burns, W. J. & Clemen, R. T. 1993. Covariance structure models and influence diagrams. Management Science, 39, 816-834.

Clarke, L. 1989, Acceptable risk: making decisions in a toxic environment. University of California Press, Berkeley, CA.

Cohrssen, J. J. & Covello, V. T. 1989. Risk analysis: a guide to principles and methods for analyzing health and environmental risks. Washington, D.C.: U.S. Council on Environmental Quality.

Covello, V. T. 1983, The perception of technological risks: a literature review. Technological Forecasting and Social Change, 23, 285-297.

Covello, V. T., McCallum, D. B. & Pavlova, M. 1989. Effective risk communication. The role and responsibility of government and non-government organizations. New York: Plenum.

Covello, V. T., & Merkhofer, M. W. 1993. Risk assessment methods: approaches for assessing health and environmental risk. New York: Plenum.

Cvetkovich, G. & Earle, T. C. 1991. Risk and culture. Special Issue, Journal of Cross-cultural Psychology.

Cvetkovich, G. & Earle, T. C. 1992. Environmental hazards and the public. Journal of Social Issues, 48, 1-20.

Daele, W. van den 1993. Backgrounds to the perception of risks in genetic engineering: concepts of nature and the semantics of risk. In: Bayerische-Rueck (Ed.): Risk is a construct, München: Knesebeck, 157-178.

Dake, K. 1991. Orienting dispositions in the perception of risk - An analysis of contemporary worldviews and cultural biases. Journal of Cross-cultural Psychology, 22, 61-82.

Dake, K. 1992. Myths of nature: Culture and the social construction of risk. Journal of Social issues, 48, 21-37.

Douglas, M. 1966. Purity and Danger: Concepts of Pollution of Taboo. Routledge and Kegan Paul, London.

Douglas, M. 1985. Risk acceptability according to the social sciences. Russell Sage Foundation, New York.

Douglas, M. & Wildavsky, A. 1982. Risk and culture: An essay on the selection of technical and environmental dangers. Berkeley: University of California Press.

Drottz-Sjöberg, B. 1991. Risk perceptions and definitions of risk. In: Drottz-Sjöberg, B. (Ed.): Perceptions of Risk, Stockholm: Center for Risk Research, 163-214.

Drottz-Sjöberg, B. & Sjöberg, L. 1991. Adolescents attitudes to nuclear power and radioactive wastes. In: Drottz-Sjöberg, B. (Ed.): Perception of Risk, Stockholm: Center for Risk Research, 109-135.

Durkheim, E. 1933, The division of labor in society. The Free Press, New York (orig. 1893).

Earle, T. C. & Lindell, M. K. 1984. Public perceptions of industrial risks: A free-response approach. In: Waller, R. A. & Covello, V. T. (Eds.): Low-probability / high-consequence risk analysis, New York: Plenum Press, 531-550.

Earle, T. C. & Cvetkovich, G. 1988. Plattitudes and comparisons: A critique of current (wrong) directions in risk communication. Proceedings of the Annual Meeting of the Society of Risk Analysis.

Edwards, W. & Winterfeldt, D. v. 1987. Public values in risk debates. Risk Analysis, 7, 141-158.

Eiser, J. R., Hannover, B., Mann, L., Morin, M., Pligt, J. v. d. & Webley, P. 1990. Nuclear attitudes after Chernobyl: A cross-national study. Journal of Environmental Psychology, 10, 101-110.

Englander, T., Farago, K., Slovic, P. & Fischhoff, B. 1986. A comparative analysis of risk perception in Hungary and the United States. Journal of Social Behavior, 1, 55-66.

Evers, A. & Nowotny, H. 1987. Über den Umgang mit Unsicherheit. Die Entdeckung der Gestaltbarkeit von Gesellschaft. Frankfurt/Main: Suhrkamp.

Femers, S. 1993. Information über technische Risiken. Bern: Lang.

Festinger, L. 1957. A theory of cognitive dissonance. Stanford University Press, Stanford CA.

Fiorino, D. J. 1989. Technical and democratic values in risk analysis. Risk Analysis, 9, 293-299.

Fischer, G. W., Morgan, M. G., Fischhoff, B., Nair, I. & Lave, L. B. 1991. What risks are people concerned about? Risk Analysis, 11, 303-315.

Fischhoff, B. 1989. Risk: a guide to controversy. In: National Research Council: Improving risk communication, Washington, D.C.: National Academy Press, Appendix C.

Fischhoff, B. 1994. Acceptable risk: a conceptual proposal. Risk: Health, Safety & Environment, 5, 1-28.

Fischhoff, B. 1996. Risk perception and communication unplugged: twenty years of process. Risk Analysis, 15, 137-145

Fischhoff, B., Slovic, P., Lichtenstein, S., READ, S. & COMBS, B. 1978. How safe is safe enough? A psychometric study of attitudes toward technological risk and benefits. Policy Science, 9, 127-152.

Fischhoff, B., Slovic, P. & Lichtenstein, S. 1981. Lay foibles and expert fables in judgements about risk. In: O'Riordan, T. & Turner, R. K. (Eds.): Progress in resource management and environmental planning, Chichester: Wiley, 161-202.

Fischhoff, B. Lichtenstein, S., Slovic, P., Derby, S. L. and Keeney, R. L. 1981. Acceptable risk. Cambridge University Press, Cambridge, MA.

Fischhoff, B., Watson, S. R. & Hope, C. 1984. Defining risk. Policy Sciences, 17, 123-129.

Fischhoff, B., Bostrom, A. & Quadrel, M. J. 1993. Risk perception and communication. Annual Review of Public Health, 14, 183-203.

Freudenburg, W. R. 1988. Perceived risk, real risk: social science and the art of probabilistic risk assessment. Science, 242, 44-49.

Gardner, G. T. & Gould, L. C. 1989. Public perceptions of the risk and benefits of technology. Risk Analysis, 9, 2, 225-243.

Goszcynska, M., Tyszka, T. & Slovic, P. 1991. Risk perception in Poland: A comparison with three other countries. Journal of Behavioral Decision Making, 4, 179-193.

Gould, L.C., Gardner, G.T. De Luca, D. R. Tiemann, A., Doob, L. W, & Stolwjik, J.A.J. 1988. Perceptions of technological risks and benefits. Russel Sage Foundation, New York.

Gregory, R. & Mendelsohn, R. 1993. Perceived risk, dread and benefits. Risk Analysis, 13, 259-264.

Gregory, R. & Lichtenstein, S. 1994. A hint of risk: Tradeoffs between quantitative and qualitative risk factors. Risk Analysis, 14, 199-207.

Guerlin, B. 1991. Psychological perspectives on risk perception and response. In: Handmer, J., Dutton, B., Guerin, B. & Smithson, M. (Eds.): New perspectives on uncertainty and risk, Canberra: ANU/ CRES, 79-86.

Haller, M. 1990. Risiko-Management und Risiko-Dialog. In: Schuez, M. (Hg.): Risiko und Wagnis. Die Herausforderung der industriellen Welt, Pfullingen: Neske, 229-256.

Handmer, J., Dutton, B., Guerin, B. & Smithson, M. 1991. New perspectives on uncertainty and risk. Canberra: ANU.

Hartenian, L. S., Bobko, P. & Berger, P. K. 1993. An empirical validation of bipolar risk perception scaling methods. Journal of Applied Social Psychology, 23, 335-351.

Heimer, C. A. 1988. Social structure, psychology, and the estimation of risk. Annual Review of Sociology, 14, 491-519.

Hillgartner, S. 1992. The social construction of risk objects: Or, how to pry open networks of risk. In: Short, J. F. & Clarke, L. (Eds.): Organizations, Uncertainties and Risk, Westview, Boulder, Co., pp. 39-53.

Hinman, G. W., Rosa, E. A., Kleinhesselink, R. R. & Lowinger, T. C. 1993. Perception of nuclear and other risks in Japan and the United States. Risk Analysis, 13, 449-456.

HMSO, 1988. The tolerability of risk from nuclear power stations. Health and Safety Executive, London.

Hoefer, M. P. & Raju, V. R. 1989. Technological risk perception: similarities and dissimilarities in french and american samples. In: Society for Risk Analysis (Ed.): Proceedings, Annual Meeting 1989, 1-6.

Hohenemser, C., Kates, R. W. & Slovic, P. 1983. The nature of technological hazards. Science, 220, 378-384.

Holtgrave, D. & Weber, E. U. 1993. Dimensions of risk perception for financial and health risks. Risk Analysis, 13, 553-558.

Horvath, P. & Zuckerman, M. 1992. Sensation seeking, risk appraisal and behavior. Personality and Individual Differences, 14, 41-53.

IEC, 1993. Guidelines for risk analysis of technological systems. Report IEC-CD (Sec) 381 issued by the Technical Committee QMS/23. European Community, Brussels.

Jianguang, Z. 1994. Environmental hazards in the Chinese public's eyes. Risk Analysis, 14, 163-169.

Johnson, B. B. 1987. The environmentalist movement and grip/group analysis: A modest critique. In: Covello, V. T. and Johnson, B. B. (Eds.): The social and cultural construction of risk, Dordrecht: Reidel, pp. 147-178.

Johnson, B. B. & Covello, V. T. (Eds.) 1987. The social and cultural construction of risk. Dordrecht: Reidel.

Johnson, E. J. & Tversky, A. 1984. Representations and perceptions of risk. Journal of Experimental Psychology (General), 113, 55-70.

Jungermann, H., Schuetz, H. & Thuering, M. 1988. Mental models in risk assessment: Informing people about drugs. Risk Analysis, 8, 147-155.

Jungermann, H., Kasperson, R. M. & Wiedemann, P. M. (Eds.) 1988. Themes and tasks of risk communication (Proceedings of an International Workshop at the KFA Jülich). Jülich: KFA-Report.

Jungermann, H., Rohrmann, B. & Wiedemann, P. 1991. Risikokontroversen - Konzepte, Konflikte, Kommunikation. Berlin etc.: Springer.

Jungermann, H. & Slovic, P. 1993. Characteristics of individual risk perception. In: Bayerische-Rueck (Ed.): Risk - a construct. München: Knesebeck, 85-102.

Kahnemann, D. & Tversky, A. 1979. Prospect theory: an analysis of decision under risk. Econometrica , 47, 263-291.

Kasperson, R. E. & Kasperson, J. X. 1983. Determining the acceptability of risk: ethical and policy issues. In: Rogers, J. T. and Bates, D. V. (Eds.): Assessment and perception of risk to human health, Conference Proceedings, Royal Society of Canada, Ottawa, pp. 135-155.

Kasperson, R. E. & Kasperson, J. X. 1987. Nuclear risk analysis in comparative perspective. Allen and Unwin, Winchester.

Kasperson, R. E., Renn, O., Slovic, P., Brown, H. S. Eemel, J., Goble, R., Kasperson, J. X. & Ratick, S. 1988. The social amplification of risk: a conceptual framework. Risk Analysis, 8, 177-187.

Kasperson, R. E. & Stallen, P. M. (Eds.) 1990. Communicating risks to the public. Dordrecht: Kluwer.

Kemp, R. 1993. Risk perception: the assessment of risks by experts and by lay people - a rational comparison? In: Bayerische-Rueck (Ed.): Risk - a construct, München: Knesebeck, 103-118.

Keown, C. F. 1989. Risk perception of Hong Kongese versus Americans. Risk Analysis, 9, 401-407.

Klages, H. 1984. Wertorientierungen im Wandel. Rückblick, Gegenwartsanalyse, Prognose. Campus, Frankfurt/Main.

Kleinhesselink, R. R. & Rosa, E. A. 1991. Cognitive Representation of Risk Perceptions - A comparison of Japan and the United States. Journal of Cross-cultural Psychology, 22, 11-28.

Kolluru, R. V. 1995. Risk assessment and management: a unified approach. In: Kolluru, R., Bartell, S., Pitblado, R. and Stricoff, S. (Eds.): Risk assessment and management handbook for environmental, health, and safety professionals, Mc-Graw-Hill: New York, pp. 1.3-1.41.

Kolluru, R. V. & Brooks, D. G. 1995. Integrated risk assessment and strategic management. In: Kolluru, R., Bartell, S., Pitblado, R. and Stricoff, S. (Eds.): Risk assessment and management handbook for environmental, health, and safety professionals, Mc-Graw-Hill: New York, pp. 2.1-2.23.

Leiss, W. (Ed.) 1990. Prospects and problems in risk communication. Waterloo: University of Waterloo Press.

Lichtenstein, S., Slovic, P., Fischhoff, B. & Combs, B. 1978. Judged frequency of lethal events. Journal of Experimental Psychology: Human Learning and Memory, 4, 551-578.

Lipset, S. M. & Schneider, W. 1983. The confidence gap, business, labor, and government in the public mind. The Free Press, New York.

Lopes, L. L. 1983. Some thoughts on the psychological concept of risk. Journal of Experimental Psychology: Human Perception and Performance, 9, 137-144.

Lopes, L. L. 1992. Risk perception and the perceived public. In: Bromley, D. W. & Segerson, K. (Eds.): The social response to environmental risk, Dordrecht: Kluwer, 57-74.

Luce, R. D. & Weber, E. U. 1986. An axiomatic theory of conjoint, expected risk. Journal of Mathematical Psychology , 30, 188-205.

Luhmann, N. 1973. Vertrauen: Ein Mechanismus der Reduktion sozialer Komplexität. Second Edition, Enke, Stuttgart.

Luhmann, N. 1980. Trust and power. Wiley, New York.

Luhmann, N. 1986. Ökologische Kommunikation. Westdeutscher Verlag, Opladen.

Luhmann, N. 1990. Technology, environment, and social risk: a systems perspective. Industrial Crisis Quarterly, 4, 223-231.

Luhmann, N. 1993. Die Moral des Risikos und das Risiko der Moral. In: Bechmann, G. (Hg.): Technik und Gesellschaft, Opladen: Westdeutscher Verlag.

Macgregor, D. 1991. Worry over technological activities and life concerns. Risk Analysis, 11, 315-325.

Machlis, G. & Rosa, E. A. 1990. Desired risk: Broadening the social amplification of risk framework. Risk Analysis, 10, 161-168.

Markowitz, J. 1991. Kommunikation über Risiken: Eine Problemskizze. Manuscript, University of Bielefeld, Bielefeld.

Marris, C., Langford, I. & O'Riordan, T. 1996. Integrating sociological and psychological approaches to public perceptions on environmental risks: Detailed results from a questionnaire survey. Norwich: University of East Anglia.

McDaniels, T. L. & Gregory, R. S. 1991. A framework for structuring cross-cultural research in risk and decision making. Journal of Cross-cultural Psychology, 22, 103-128.

Mechitov, A. I. & Reblik, S. B. 1990. Studies of risk and safety perception in the USSR. In: Borcherding, K., Larichev, O. J. & Mesick, D. M. (Eds.): Contemporary issues in decision making, Amsterdam: North-Holland, 261-270.

Morgan, M. G. 1990. Choosing and managing technology-induced risks. In: Glickman, T. S. & Gough, M. (Eds.), Readings in risk, Resources for the Future, Washington D.C., pp. 5-15.

Mullet, E., Duquesnoy, C., Raiff, P., Fahrasmane, R. & Namur, E. 1993. The evaluative factor of risk perception. Journal of Applied Social Psychology, 23, 1594-1605.

National Research Council, Committee on the Institutional Means for Assessment of Risks to Public Health 1983. Risk assessment in the Federal Government: managing the process. National Academy of Sciences, National Academy Press, Washington D.C.

Nelkin, D. 1982. Blunders in the business of risk. Nature, 298, 775-776.

Nyland, L. G. 1993. Risk perception in Brazil and Sweden. Stockholm School of Economics: Centre for Risk Research.

O'Riordan, T. 1983. The cognitive and political dimensions of risk analysis. Journal of Environmental Psychology, 3, 345-354.

Otway, H. & Thomas, K. 1982. Reflections on risk perception and policy. Risk Analysis, 2, 69-82.

Parsons, T. E. 1960. Pattern variables revisited. American Sociological Review, 25, 467-483.

Perusse, M. 1980. Dimensions of perception and recognition of danger. Birmingham: Diss., Univ. of Aston. (cited from Slovic, P. 1991)

Peters, E. & Slovic, P. 1996. The role of affect and worldviews as orienting dispositions in the perception and acceptance of nuclear power. Journal of Applied Social Psychology, 26, 1427-1453.

Peters, H. P. 1991: Warner oder Angstmacher? Thema Risikokommunikation. In: K. Mertens; S. J. Schmidt & S. Weischenberg (Hg.) Funkkolleg 'Medien und Kommunikation' Konstruktionen von Wirklichkeit. Studienbrief 9. Weinheim & Basel: Beltz, 74-108.

Peters, H. P. 1999: Das Bedürfnis nach Kontrolle der Gentechnik und das Vertrauen in wissenschaftliche Experten. In: J. Hampel & O. Renn (Hg.): Gentechnik in der Öffentlichkeit. Wahrnehmung und Bewertung einer umstrittenen Technologie. Frankfurt/Main: Campus, 225-245.

Petty, R. E & Cacioppo, E. 1986. The elaboration likelihood model of persuasion. Advances in Experimental Social Psychology, 19, 123-205.

Pidgeon, N., Hood, C., Jones, D., Turner, B. & Gibson, R. 1992. Risk perception. In: The Royal Society: Risk Analysis, perception and management, London, 89-134.

Plough, A. & Krimsky, S. 1987. The emergence of risk communication studies in science and political context. Science, Technology and Human Values, 12, 4-10.

Pollatsek, A. & Tversky, A. 1970. A theory of risk. Journal of Mathematical Psychology, 7, 540-553.

Puy, A. & Aragones, J. I. 1994. Risk perception and the management of emergencies (Paper presented at the 23rd International Congress of Applied Psychology). Madrid: Dept. of Psychology, Univ. of Madrid.

Rayner, S. 1987. Risk and relativism in science for policy. In: Covello, V. T. & Johnson, B. B. (Eds.), The social and cultural construction of risk, Reidel, Dordrecht, pp. 5-23.

Rayner, S. 1992. Cultural theory and risk analysis. In: Krimsky, S. & Golding, D. (Eds.): Social theories of risk, Westport: Praeger.

Rayner, S. & Cantor, R. 1987. How fair is safe enough? The cultural approach to societal technology choice. Risk Analysis, 7(1), 3-13.

Renn, O. 1983. Technology, risk and public perception. Angewandte Systemanalyse, 4, 50-65.

Renn, O. 1986. Risk perception: a systematic review of concepts and research results. In: Miller, C. T., Kleindorfer, R. P. & Munn, R. E. (Eds.): Conceptual trends and implications for risk research, Laxenburg: International Inst. for Applied System Analysis, 257-286.

Renn, O. 1990a. Risk perception and risk management. Part 1: The intuitive mechanism of risk perceptions. Risk Abstracts 7, 1-9.

Renn, O. 1990b. Risk perception and risk management. Part 2: From risk perception to risk management. Risk Abstracts, 8, 1-10.

Renn, O. 1991. Risk communication and the social amplification of risk. In: Kasperson, R. E. and Stallen, P. J. (Eds.): Communicating risk to the public, Dodrecht: Kluwer, pp. 287-324.

Renn, O. 1992a. Concepts of risk: a classification. In: Krimsky, S. & Golding, D. (Eds.), Social theories of risk, Praeger, Westport, CT, pp. 53-79.

Renn, O .1992b. The social arena concept of risk debates. In: Krimsky, S. & Golding, D. (Eds.), Social theories of risk, Praeger, Westport, CT, pp. 170-197.

Renn, O. 1992c. Risk communication: towards a rational discourse with the public. Journal of Hazardous Materials, 29, 465-519.

Renn, O. 1993. Technik und gesellschaftliche Akzeptanz: Herausforderungen der Technikfolgenabschaetzung. GAIA, 2, 67- 83.

Renn, O. 1995. Style of using scientific expertise: a comparative framework. Science and Public Policy, 22, 147-156.

Renn, O. & Swaton, E. 1984. Psychological and sociological approaches to study risk perception. Environment International, 10, 557-575.

Renn, O. & Levine, D. 1991. Trust and credibility in risk communication. In: Kasperson, R. & Stallen, P. J. (Eds.): Communicating risk to the public, Dordrecht: Kluwer, pp. 175-218.

Renn, O. Burns, W. Kasperson, R. E. Kasperson, J.X & Slovic, P. 1992. The social amplification of risk: theoretical foundations and empirical application. Social Issues, 48, No. 4, Special Issue: Public Responses to Environmental Hazards, 137-160.

Rohrmann, B. 1991. Akteure der Risiko-Kommunikation. In: Jungermann, H., Rohrmann, B. & Wiedemann, P. M. (Hg.): Risikokontroversen - Konzepte, Konflikte, Kommunikation, Berlin etc.: Springer, 355-371.

Rohrmann, B. 1991. Psychologische Risikoforschung. In: Frey, D. (Hg.): Bericht über den 37. Kongress der DGfPs, Göttingen: Hogrefe, 393-412.

Rohrmann, B. 1991. Risks and benefits of individual activities and living conditions - a cross-cultural comparison (Research Report). Hamilton/NZ: University of Waikato.

Rohrmann, B. 1992. The evaluation of risk communication effectiveness. Acta Psychologica, 81, 169-192.

Rohrmann, B. 1993. Risk management by setting standards. In: Bayerische-Rueck (Ed.): Risk as a construct, München: Knesebeck, 293-313.

Rohrmann, B. 1994. Risk perception of different societal groups: Australian findings and cross-national comparisons. Australian Journal of Psychology, 46, 150-163.

Rohrmann, B. 1995. Technological risks: Perception, evaluation, communication. In: Mechlers, R. E. & Stewart, M. G. (Eds.): Integrated risk assessment: Current practice and new directions, Rotterdam: Balkema, 7-12.

Rohrmann, B. 1998. The risk notion - epistemological and empirical considerations. In Melchers, R. (Ed.): Integrative risk assessment II. Rotterdam: Balkema.

Rohrmann, B. 1999. Risk perception research: Review and documentation. Update, Studies in Risk Communication vol. 69. Available on the WWW; URL = http://www.kfa-juelich.de/mut/hefte/heft_69.pdf. Juelich: Research Center Juelich.

Rohrmann, B. & Chen, H. 1999. Risk perception in Australia and China: an exploratory cultural study. Journal of Risk Research 2 (3), 219-241.

Rosa, E. A. 1996. Metatheoretical foundations for post-normal risk. Paper presented at the SRA-Europe Meeting, Manuscript, University of Surrey, Guildford, June 3-5, 1996 [Department of Sociology, Washington State University, Pullman, WA.]

Ross, L.D. 1977. The intuitive psychologist and his shortcomings: distortions in the attribution process. In: Berkowitz, L. (Ed.): Advances in experimental social psychology, Vol.10, Random House, New York, pp. 173-220.

Schuez, M. (Hg.) 1990. Risiko und Wagnis. Die Herausforderung der industriellen Welt. Pfullingen: Neske.

Schuetz, H. in prep. Die Struktur von Risikowahrnehmung - Anwendung von Strukturgleichungsmodellen. Jülich: KFA/MUT.

Schwarz, M. & Thompson, M. 1990. Divided we stand: redefining politics, technology, and social choice. University of Pennsylvania Press, Philadelphia.

Short, J. F. 1984. The social fabric of risk: toward the social transformation of risk analysisAmerican Sociological Review, 9, 711-725.

Short, J. F. 1989. On defining, describing, and explaining elephants (and reactions to them): hazards, disasters, and risk analysis. Mass Emergencies and Disasters, 7, 397-418.

Sjöberg, L. (Ed.) 1987. Risk and society. London: Allen & Unwin.

Sjöberg, L. 1995. How cognitive is risk perception? A discussion of the psychometric and cultural theory approaches. Paper presented to the 4th European Congress of Psychology, Athens, 2-7 July, 1995. Center for Risk Research, Stockholm School of Economics.

Sjöberg, L. 1996. A discussion of the limitations of the psychometric and cultural theory approaches to risk perception. Radiation Protection Dosimetry, 68, 219-225.

Sjöberg, L. 1997. Explaining risk perception: An empirical evaluation of cultural theory. Risk Decision and Policy, 2, 113-130.

Slovic, P. 1987. Perception of risk. Science, 236, 280-285.

Slovic, P. 1991. Beyond numbers: A broader perspective on risk perception and risk communication. In: Mayo, D. & Hollander, R. D. (Eds.): Acceptable evidence: Science and values in risk management, New York: Oxford University Press, 48-65.

Slovic, P. 1992. Perception of risk: Reflections on the psychometric paradigm. In: Golding, D. & Krimsky, S. (Eds.): Theories of risk, London: Praeger, 117-152.

Slovic, P., Fischhoff, B. & Lichtenstein, S. 1980. Facts and fears - understanding risk. In: Schwing, R. C. & Albers, W. A. (Eds.): Societal risk assessment, New York: Plenum, 181-218.

Slovic, P. Fischhoff, B. & Lichtenstein, S. 1981. Perceived risk: psychological factors and social implications. In: Royal Society (Ed.): Proceedings of the Royal Society, Report A376, Royal Society, London, pp. 17-34.

Slovic, P., Fischhoff, B. & Lichtenstein, S. 1982. Why study risk perception. Risk Analysis, 2, 83-93.

Slovic, P., Fischhoff, B. & Lichtenstein, S. 1985. Characterizing perceived risk. In: Kates, R. W., Hohenemser, C. & Kasperson, J. X. (Eds.): Perilous progress: managing the hazards of technology, Boulder: Westview, 91-125.

Slovic, P., Fischhoff, B. & Lichtenstein, S. 1986. The psychometric study of risk perceptions. In: Covello, V. T., Menkes, J. & Mumpower, J. (Eds.): Risk evaluation and management, New York: Plenum.

Sokolowska, J. & Tyszka, T. 1995. Perception and acceptance of technological and environmental risks: Why are poor countries less concerned? Risk Analysis, 15, 733-744.

Starr, C. 1969. Social benefit versus technological risk. Science, 165, 1232-1238.

Starr, C. & Whipple, C. 1991. The strategic defense initiative and nuclear proliferation from a risk analysis perspective. In: Shubik, M. (Ed.): Risk, organizations, and society, Kluwer: Dordrecht, pp. 49-61.

Stern, P. C. & Fineberg, V. 1996. Understanding risk: informed decisions in a democratic society. National Research Council, Committee on Risk Characterization (Ed.), National Academy Press, Washington, D.C.

Swaton, E. & Renn, O. 1984. Attitudes towards nuclear power: a comparison between three nations. Laxenburg: Institute for Applied System Analysis.

Teigen, K. H., Brun, W. & Slovic, P. 1988. Societal risks as seen by a Norwegian public. Journal of Behavioral Decision Making, 1, 111-130.

Thomas, K., Maurer, D. Fishbein, M., Orway, H. L., Hinkle, R. & Simpson, D. A. 1980. Comparative study of public beliefs about five energy systems. International Institute for Applied Systems Analysis (IIASA), Report 80-15, IIASA, Laxenburg, Austria.

Thompson, M. 1980. An outline of the cultural theory of risk, Working Paper of the International Institute for Applied Systems Analysis, WP-80-177, IIASA, Laxenburg, Austria.

Thompson, M., Ellis, W. & Wildavsky, A. 1990. Cultural theory. Boulder: Westview Press.
Tiemann, A. R. & Tiemann, J. J. 1985. Cognitive maps of risk and benefit perceptions. In: Whipple, C. & Covello, V. T. (Eds.): Risk analysis in the private sector, New York: Plenum Press.
Tyszka, T. & Goszcynska, M. 1993. What verbal reports say about risk perception. Acta Psychologica, 83, 53-64
Vlek, C. A. J. 1996. A multi-level, multi-stage and multi-attribute perspective on risk assessment, decision-making, and risk control. Risk, Decision, and Policy, 1, 9-31.
Vlek, C. & Stallen, P. J. 1980. Rational and personal aspects of risk. Acta Psychologica, 45, 273-300.
Vlek, C. A. J. & Stallen, P. J. 1981. Judging risks and benefits in the small and in the large. Organizational Behavior and Human Performance, 28, 235-271.
Vlek, C. & Cvetkovich, G. (Eds.) 1989. Social decision methodology for technological projects. Dordrecht: Kluwer.
Vlek, C. & Keren, G. 1992. Behavioral decision and environmental risk management: Assessment and resolution of four 'survival' dilemmas. Acta Psychologica, 80, 249-279.
Waterstone, M. (Ed.) 1991. Risk and society: The interation of science, technology and public policy. Dordrecht: Kluwer.
Wiedemann, P. M. 1993. Tabu, Sünde, Risiko: Veränderungen der gesellschaftlichen Wahrnehmung von Gefährdungen. In: Bayerische Rückversicherung (Hg.): Risiko ist ein Konstrukt. Wahrnehmungen zur Risikowahrnehmung, München: Knesebeck, pp. 43-67.
Wiedemann, P. M., Femers, S. & Hennen, L. 1991. Bürgerbeteiligung bei entsorgungswirtschaftlichen Vorhaben. Berlin: Erich Schmidt.
Wiedemann, P. M. & Kresser, R. M. 1997. Intuitive Risikobewertung - Strategien der Bewertung von Umweltrisiken. Jülich: Forschungszentrum (MUT-Texte 62).
Wildavsky, A. 1988. Searching for safety. New Jersey: Transaction Books.
Wildavsky, A. 1993. The comparative study of risk perception: a beginning. In: Bayerische Rueck: Risk - a construct, München: Knesebeck, 179-196.
Wildavsky, A. & Dake, K. 1990. Theories of risk perception: Who fears what and why? Daedalus, 119, 41-60.
Winterfeldt, D. v. & Edwards, W. 1984. Patterns of conflicts about risky technologies. Risk Analysis, 4, 55-68.
Yates, J. F. & Stone, E. 1992. The risk construct. In: Yates, J. F. (Ed.): Risk-taking behavior, Chichester: Wiley, 1-26.

2

NUCLEAR POWER AND THE PUBLIC

A COMPARATIVE STUDY OF RISK PERCEPTION IN FRANCE AND THE UNITED STATES

Paul Slovic
Decision Research, Eugene, Oregon, USA

James Flynn
Decision Research, Eugene, Oregon, USA

C. K. Mertz
Decision Research, Eugene, Oregon, USA

Marc Poumadère
Ecole Normale Supérieure, Cachan, France

Claire Mays
Institut Symlog, Cachan, France

TABLE OF CONTENTS

SUMMARY

This study is an attempt to understand attitudes, perceptions, and behaviors with respect to nuclear power and several other technological risk sources. A unique feature of the study is a comparison between public views in the United States, where nuclear energy is resisted, and France, where nuclear energy appears to be embraced (France obtains about 80% of its electricity from nuclear power).

Although the experiences of France and the U.S. with regard to nuclear energy overlap during the post World War II decades, there are a number of significant differences in timing, motivation toward adopting nuclear power, the economic context, the cultural and political milieu, regulation, and financing of the industry. We would expect these conditions to be associated with significant differences between French and American attitudes and opinions about nuclear power and related issues.

We have used the same survey, between public views in the United States and public views in France. We found, to our surprise, that concerns about nuclear power and nuclear waste as measured by the survey were high in France and were at least as great there as in the U.S. Thus, even though perception of risk is a strong predictor of attitudes toward nuclear power within both the U.S. and France, it cannot account for the different level of reliance on nuclear energy in the two countries. Further analysis of the survey data uncovered a number of differences that might be central in explaining the difference between France and the U.S. Specifically, the French:

- saw greater need for nuclear power and greater economic benefit from it;
- had greater trust in scientists, industry, and government officials who design, build, operate, and regulate nuclear power plants;
- were more likely to believe that decision-making authority should reside with the experts and government authorities, rather than with the people.

These findings point to some important differences between the workings of democracy in the U.S. and France and the effects of different "democratic models" on acceptance of risks from technology.

1 BACKGROUND AND OBJECTIVES OF THE STUDY

The history of nuclear power is dramatic. Its creation grows out of and reflects the greatest achievements of science and engineering in the 20th century. Its early years demonstrated rapid progress toward the promise of "energy too cheap to meter". By the mid-1970s its progress in the United States came to a halt, and the past two decades have witnessed the growth of strong fear of this technology and opposition to it among many members of the American public. Nuclear scientists and many others in industry and government, however, remain convinced of its safety and efficacy as an energy source, and they express great consternation and antagonism toward a public they view as "ignorant", "radiophobic" or "irrational".

The present study is an attempt to understand attitudes, perceptions, and behaviors that lie at the heart of this conflict. A unique feature of the study is a comparison between public views in the United States, where nuclear energy is resisted, and France, where nuclear energy appears to be embraced (France obtains about 80% of its electricity from nuclear power).

Although the experiences of France and the U.S. with regard to nuclear energy overlap during the post World War II decades, there are a number of significant differences in timing, motivation toward adopting nuclear power, the economic context, the cultural and political milieu, regulation, and financing of the industry. We would expect these conditions to be associated with significant differences between French and American attitudes and opinions about nuclear power and related issues.

Based on the history of the two programs we constructed a set of hypotheses about key attitudes and opinions that might differ in France and the U.S.:

- The French will exhibit greater support for nuclear power than the Americans
- The French will assess nuclear power as less risky
- The French will have greater trust and confidence in the nuclear industry
- The French will have greater confidence in science and technology
- The French will be more fatalistic about their ability to influence personal risks or public affairs
- The French will consider nuclear power as economically more necessary
- The French will have more confidence that future risks from nuclear power can be managed adequately.

An extensive survey was designed to evaluate these hypotheses and to assess many different aspects of risk perception and other social, psychological, political, and economic attitudes that we hypothesized might underlie support or opposition to nuclear power in the U.S. and France.

2 METHODOLOGY

2.1 Survey Content

The survey was designed to assess many different aspects of risk perception and other social, psychological, and economic attitudes that we hypothesized might underlie nuclear power support and opposition. A variety of question formats were used, including word associations, ratings of perceived risk, attitude and opinion questions, and questions about voluntary risk-taking activities in which the respondent has engaged. Question content was based on a general model of the way various factors might influence support for nuclear power. This model was derived from the extensive literature on risk perception and nuclear power. The main components of the survey are outlined in Table 1 and discussed below in approximately the order in which they were presented to the respondents. In this chapter, we shall focus on the results from the survey components marked with an asterisk (*).

Table 1. **OUTLINE OF THE CONTENT OF THE FRENCH AND AMERICAN SURVEY**

1. **Word associations***
   Associations to "nuclear power" and ratings of these associations on a positive/negative scale
2. **Perceived risk***
   a. Risk to the public as a whole from each of 25 hazards, including nuclear power and other radiation hazards (e.g., radon, X-rays, food irradiation, suntanning, video-display terminals)
   b. Risk to you and your family on a subset of these 25 items
3. **National problems**
   Risk from nuclear power versus other social/political/economic problems
4. **Health and environmental problems**
5. **Intuitive toxicology**
6. **Trust in experts/government/science***
7. **Risk/benefit tradeoffs**
8. **Knowledge**
9. **Nuclear power***
   a. Support versus opposition
   b. Acceptability compared to other sources of energy
   c. Attitudes/opinions/characteristics
   d. Comparison between characteristics of nuclear power risks and risks from oil and coal
10. **Equity and fairness: Nuclear power and societal decisions**
11. **Demand, scarcity, and conservation of electricity**
12. **Worldviews***

a. Fatalism	e. Technological enthusiasm/catastrophism
b. Hierarchy	f. Conservative/authoritarian ideology
c. Egalitarian view	g. Concern for future generations
d. Individualism	h. Economic growth

13. **Personal data***

a. Gender	f. Personal health
b. Age	g. Race
c. Children under 18	h. Environmental activism
d. Education	i. Income
e. Voluntary risk taking	j. Actual distance to nearest nuclear power plant
	k. Perceived distance to nearest nuclear power plant

Note. Results from survey sections marked with an asterisk (*) are presented and discussed in the present chapter.

WORD ASSOCIATIONS

The study of associations has a long history in psychology, going back to Galton (1880), Wundt (1883) and Freud (1924). More recently, word associations have been found useful in revealing important aspects of the ways people perceive risks (see, e.g., Jenkins-Smith, 1993; Slovic, Kraus, Lappe, Letzel, & Malmfors, 1989; Slovic, Flynn, & Layman, 1991; Slovic, Layman, &

Flynn, 1991). The first question in the survey asked respondents to indicate the first word or image that came to mind when they heard the words "nuclear power." They were prompted to produce up to three words or images. These associations were elicited at the beginning of the interview so that respondents would not be influenced by the content of other questions. The respondents were then asked to rate the affective quality of their images to nuclear power as very negative, negative, neutral, positive, or very positive.

PERCEIVED RISK

In the next segment of the interview, respondents were asked to indicate the degree of health risk they associated with each of the 25 hazards presented to them. These 25 items covered a wide range of hazards, including risks from technology (e.g., nuclear power, high-voltage power lines), lifestyle (e.g., AIDS, suntanning, cigarette smoking), pollution (e.g., indoor air quality, chemical pollution in the environment), common substances (e.g., bacteria in food), and natural hazards (e.g., storms and floods). The array of items was selected to include many of the hazards that have recently been of concern to the public, medical community, or government agencies. In addition to nuclear power, there were five other items pertaining to various forms of radiation risk (e.g., radon, X-rays, food irradiation, suntanning, video display terminals).

Each item was rated in terms of the health risk posed to the American (or French) public as a whole. The possible responses were "almost no health risk," "slight health risk," "moderate health risk," and "high health risk." In addition, for six of the items thought to pose risks to respondents and their families that might be quite different from the risks to the general public (e.g., street drugs), ratings were also obtained for "the health risk to you and your family."

TRUST IN EXPERTS, GOVERNMENT, AND SCIENCE

Trust has been found to play a central role in determining public perceptions of risk from nuclear power and public opposition to nuclear facilities. Six items were asked to elicit respondents' level of trust in experts, government, and scientists to manage health and technological risks (e.g., "Decisions about health risks should be left to the experts").

NUCLEAR POWER

Five items dealt with support or opposition to nuclear power under differing conditions. Two of these statements asked about attitudes toward building a new nuclear power plant in the face of a shortage of electricity or to avoid importing energy from other countries. Another asked about building new nuclear power plants to reduce health and environmental damage from coal and oil. A third statement asserted we should stop using nuclear power plants because we have no way to safely store the radioactive wastes. The final statement asked for a response to the prospect of a new and safer generation of nuclear power plants.

A second set of questions asked the respondents to rate nuclear power and six other ways to produce electricity. The four-point scale presented to the respondents was from "not at all acceptable" to "slightly acceptable" to "moderately acceptable" and finally to "very acceptable."

The alternative sources of energy production were oil, solar power, natural gas, hydroelectric power, wind power, and coal.

People hold a wide variety of opinions about the characteristics of nuclear power. Some of these characteristics are hypothesized to have an influence on the decision to support or oppose nuclear power. Ten statements were evaluated by respondents on the "strongly disagree" to "strongly agree" scale. These statements presented nuclear power as a technological achievement worthy of pride, as essential to national security and the economy, as a contributor to nuclear-weapons production, as immoral for the risks it imposes on future generations, and as likely to stigmatize host communities. One item asked about whether scientists agree on the acceptability of nuclear-power risks, and another asked whether we should develop alternatives to nuclear power and fossil fuel plants.

The risks from nuclear power have been evaluated in a number of previous studies as being poorly understood (by the public and by scientists), uncontrollable, and catastrophic (Slovic, 1987). The present survey presented four statements asking for evaluations of nuclear power as understood by science and the public, controllable by science, and presenting catastrophic risks. Four similar statements asked for evaluations of coal and oil risks on the same characteristics.

WORLDVIEWS

Over the past decade, evidence has been accumulating regarding the importance of general dispositions or "worldviews" in determining an individual's perceptions of risk (Buss, Craik, & Dake, 1986; Dake, 1991; Jasper, 1990). The survey contained a number of statements designed to measure the following worldviews:

- Fatalism/control (e.g., "It's no use worrying about public affairs; I can't do anything about them anyway")
- Hierarchy (e.g., "We need to pull together and support the energy choices our government has made")
- Egalitarian (e.g., "If people in this country were treated equally, we would have fewer problems")
- Individualism (e.g., "In a fair system people with more ability should earn more")
- Technological enthusiasm/catastrophism (e.g., "A high-technology society is important for improving our health and well-being")
- Conservative/authoritarian (e.g., "I am in favor of capital punishment")
- Future generations (e.g., "Our technologies might impose risks on future generations, but I believe future generations will be able to take care of themselves")
- Economic growth (e.g., "Continued economic growth is necessary to improve our quality of life").

Some of the other items described earlier can also be considered as tapping worldviews. For example, items asking about trust in experts and government officials could be considered measures of the hierarchy worldview.

PERSONAL AND DEMOGRAPHIC CHARACTERISTICS

The final section of the survey elicited information pertaining to the personal background of the respondent. In addition to standard demographic information (e.g., gender, age, education, income), respondents were asked about their health status, voluntary risk taking, political orientation, their degree of political activism, and their environmental activism (e.g., "Have you purchased a higher priced product because it was better for your health or environmentally friendly in the past year?").

2.2 Administration of the Survey

The survey was administered in both the United States and France. In each country a representative sample of the adult population was interviewed by telephone. Respondents were chosen based on a random-digit dialing process combined with recruiting the person in the household who was 18 years of age or older and had the most recent birth date. The same questionnaire was used in both countries with only slight necessary modifications (e.g., French respondents were asked to rate the risk to the French public; Americans to the American public). The instrument included 155 items, and the average length of the interview was approximately 30 minutes in both countries.

FRANCE

Interviews for the French sample were conducted by Brule Ville Associes from November 26 to December 12, 1992; 1550 interviews were completed with a response rate of 49.7%. The sample included 729 (47.0%) men and 821 (53.0%) women with a mean age of 42.2 years.

UNITED STATES

The U.S. sample consisted of 1512 English-speaking persons. The interviews were conducted by O'Neil Associates between November 21, 1992, and January 16, 1993. The response rate was 50.7%. The sample included 729 men (48.2%) and 783 women (51.8%) with a mean age of 39.7 years.

3 RESULTS: UNIVARIATE ANALYSES

3.1 Perceptions of Risk and Major National Problems

RISK TO THE AMERICAN PUBLIC

Figure 1 displays the perceived risk to the American public as a whole for nuclear power, nuclear waste, and 23 other items. Cigarette smoking, street drugs, AIDS, and nuclear waste elicited the greatest percentage of responses in the "high-risk" category, with at least 60% high-risk responses. Other items eliciting over 40% in the high-risk category were stress, chemical pollu-

Figure 1. **PERCEIVED HEALTH RISKS TO AMERICAN PUBLIC**

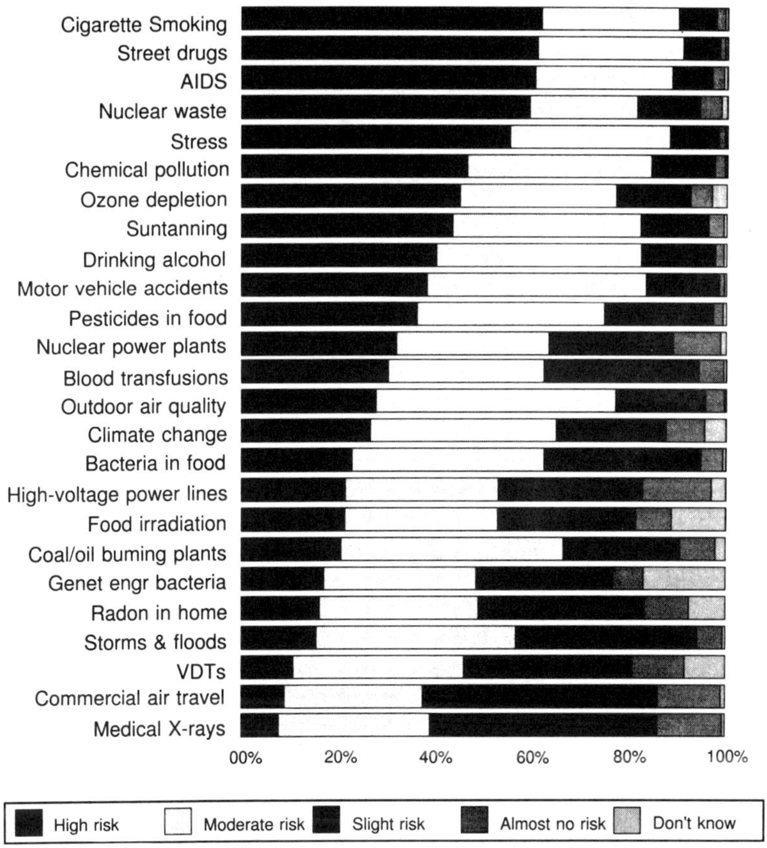

tion, depletion of the ozone, suntanning, and drinking alcohol. Commercial air travel and medical X-rays received the lowest percentage of high-risk responses. One hazard that experts might see as somewhat serious, bacteria in food, was rated moderately low in risk to health.

Chemical risks from ozone, street drugs, chemical pollution, pesticides, and alcohol were rated high in risk. Radiation hazards associated with industry (nuclear waste and nuclear power) were seen as more risky than radiation hazards associated with medicine (X-rays). The relatively low perceived risks associated with medical uses of radiation replicates earlier findings in Canada (Slovic, Flynn, Mertz, & Mullican, 1993; Slovic et al., 1989) and elsewhere (Slovic, 1990) and may reflect the influences on risk perceptions of perceived benefits, familiarity, and trust, all of which are higher for medical treatments and the medical establishment than for nuclear power and the nuclear industry.

Perceived risks due to climate change and coal and oil power plants were seen as lower in magnitude compared to risks from chemical pollution of the environment and depletion of ozone.

Figure 2. **PERCEIVED RISK TO "YOU AND YOUR FAMILY" AND TO THE "AMERICAN PUBLIC AS A WHOLE". BARS REPRESENT PERCENTAGE OF "HIGH-RISK" RESPONSES**

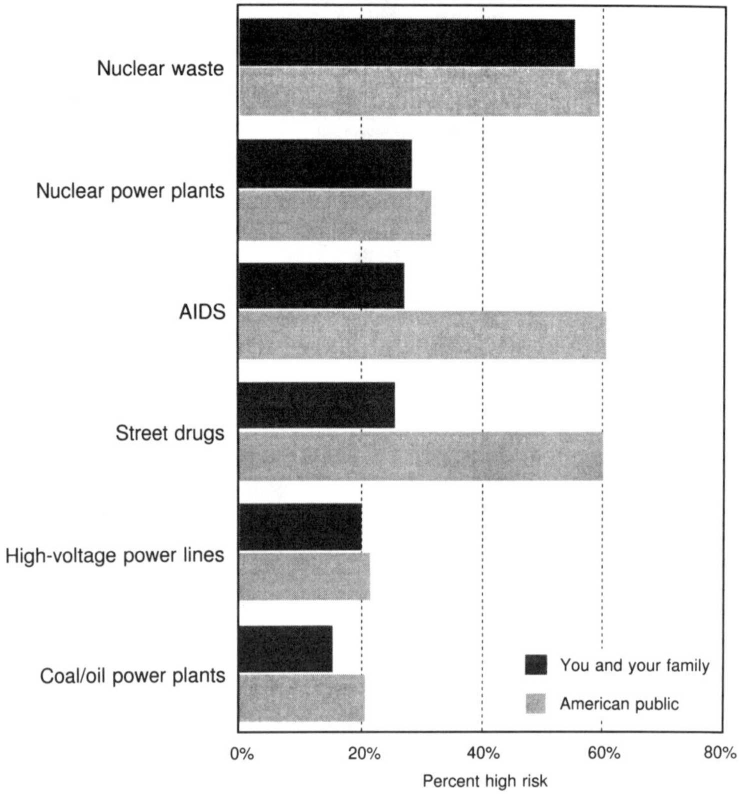

Nuclear waste was seen as a more serious risk than nuclear power, a finding also observed in previous studies (Slovic et al., 1993).

RISK TO AMERICAN RESPONDENTS AND THEIR FAMILIES

Figure 2 compares the percentage of high-risk responses when respondents were considering the health risk "to you and your family" with the percentage for the "American public as a whole" for each of the six items. Note that nuclear waste received more high-risk response evaluations than any other items when the reference was to personal or familial risk. Nuclear power ranked second in this respect. In every case, there were more high-risk judgments in reference to the American public although, in some cases, the differences were small (e.g., nuclear power plants, nuclear waste, high-voltage power lines). However, for two items, AIDS and street drugs, the difference between personal and societal risk perception was quite large.

Drottz-Sjöberg (1993) demonstrated that the difference between personal and societal risk perceptions is linked to perceived personal control over the hazard. It is noteworthy that our respondents judged the risk to themselves and their families from nuclear power to be about as great as the risk to the public as a whole. Slovic et al. (1993) observed the same finding in Canada. All but one of Canada's nuclear power plants are in Ontario, yet people everywhere in Canada, even thousands of miles from the nearest reactor, perceived the risks to themselves and their families to be as great as the risks to the Canadian public.

Figure 3. **PERCEIVED HEALTH RISKS TO AMERICAN PUBLIC BY GENDER: DIFFERENCE BETWEEN MALES AND FEMALES**

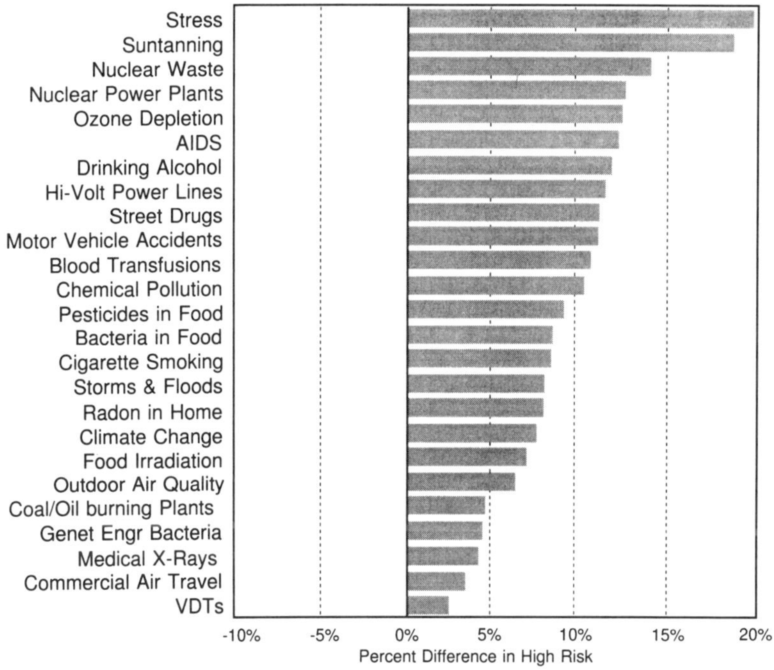

Note: Difference is percent female high-risk responses minus percent male high-risk responses.

GENDER DIFFERENCES: U.S. SAMPLE

Perceived risk to the American public as a whole was examined by gender. Sizable differences were found between perceptions of men and women. These differences are illustrated in Figure 3. Women were more likely to rate a risk as high-risk for all 25 hazards. In many instances, the differences were quite large - almost 20% for stress and suntanning. Other items exhibiting more than a 10% difference in high-risk responses were nuclear waste, nuclear power plants, ozone depletion, AIDS, drinking alcohol, high-voltage power lines, streets drugs, motor vehicle accidents, blood transfusions, and chemical pollution. Items for which women had relatively less excess concern (when compared to men) included video display terminals, commercial air travel, medical X-rays, and genetically engineered bacteria.

Figure 4. **PERCEIVED HEALTH RISKS TO AMERICAN PUBLIC BY AGE: DIFFERENCE BETWEEN 55+ AND UNDER 30 AGE GROUPS**

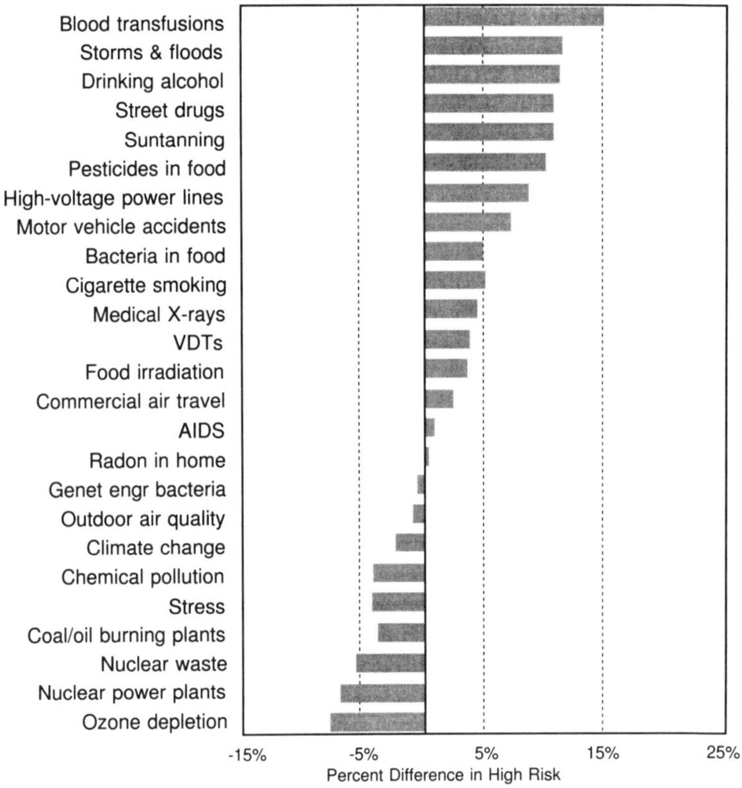

Note: Percent difference is percent 55+ high-risk responses minus under-30 high-risk responses.

AGE DIFFERENCES: U.S. SAMPLE

Figure 4 displays the difference in high-risk responses between respondents age 30 or less and age 55 or more. In general, older persons are more likely to rate a health risk as high. This tendency was particularly evident for blood transfusions, storms and floods, drinking alcoholic beverages, street drugs, suntanning, and pesticides in food. Younger respondents exhibited slightly higher perceived risk than the older group for nuclear power plants, nuclear waste, and various forms of chemical pollution (including ozone depletion and coal and oil burning power plants). Items for which there were relatively little differences between the older and younger age groups included AIDS, radon in home, genetically engineered bacteria, and outdoor air quality.

Figure 5. **PERCEIVED HEALTH RISKS TO AMERICAN PUBLIC BY EDUCATION: DIFFERENCE BETWEEN SCHOOL AND COLLEGE-EDUCATED RESPONDENTS**

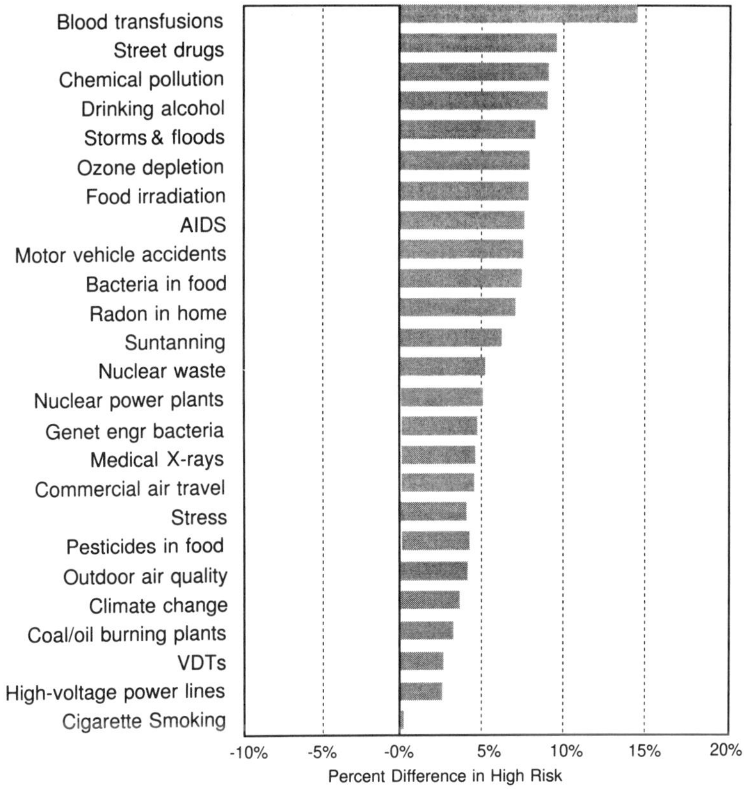

Note: Percent difference is percent high school educated high-risk responses minus percent college-educated high-risk responses.

EDUCATION DIFFERENCES: U.S. SAMPLE

Respondents with high school or less education were consistently more likely than college educated respondents to rate a risk as "high" (see Figure 5). People with less formal education were particularly likely to see blood transfusions as a high risk - with nearly 15% more high-risk responses. There were 14 additional items for which they recorded over 5% more high-risk responses than college educated respondents, including street drugs, various forms of chemical pollution, drinking alcohol, storms and floods, AIDS, nuclear waste, and nuclear power plants. In general, these differences were smaller than gender differences.

Figure 6. **PERCEIVED HEALTH RISKS TO FRENCH PUBLIC**

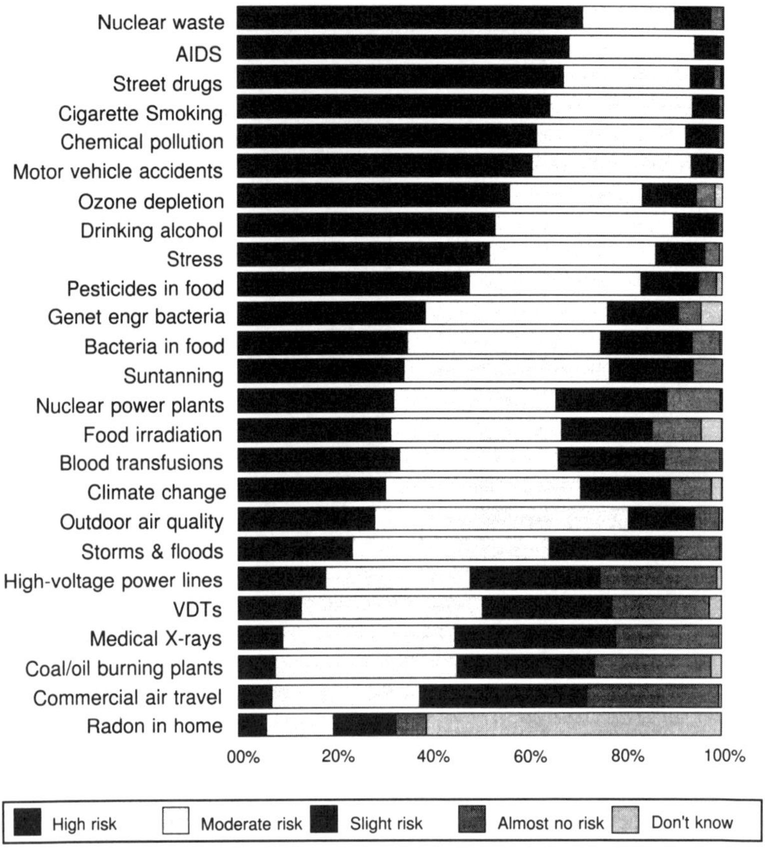

RISK TO THE FRENCH PUBLIC

Figure 6 shows the perceived risk to the French public as a whole for 25 items. Nuclear waste elicited the greatest percentage of high-risk responses followed by AIDS, street drugs, and cigarette smoking. Other items eliciting more than 40% in the high-risk category were chemical pollution, motor vehicle accidents, ozone depletion, drinking alcoholic beverages, stress, and pesticides in food. Radon in the home elicited the lowest percentage of high-risk responses. Radiation hazards associated with nuclear waste and nuclear power were seen as more risky than radiation hazards associated with medical X-rays. As in the U.S., nuclear waste was seen as more hazardous than nuclear power plants.

Figure 7. **PERCEIVED RISK TO "YOU AND YOUR FAMILY" AND TO THE "FRENCH PUBLIC AS A WHOLE": BARS REPRESENT PERCENTAGE OF "HIGH-RISK" RESPONSES**

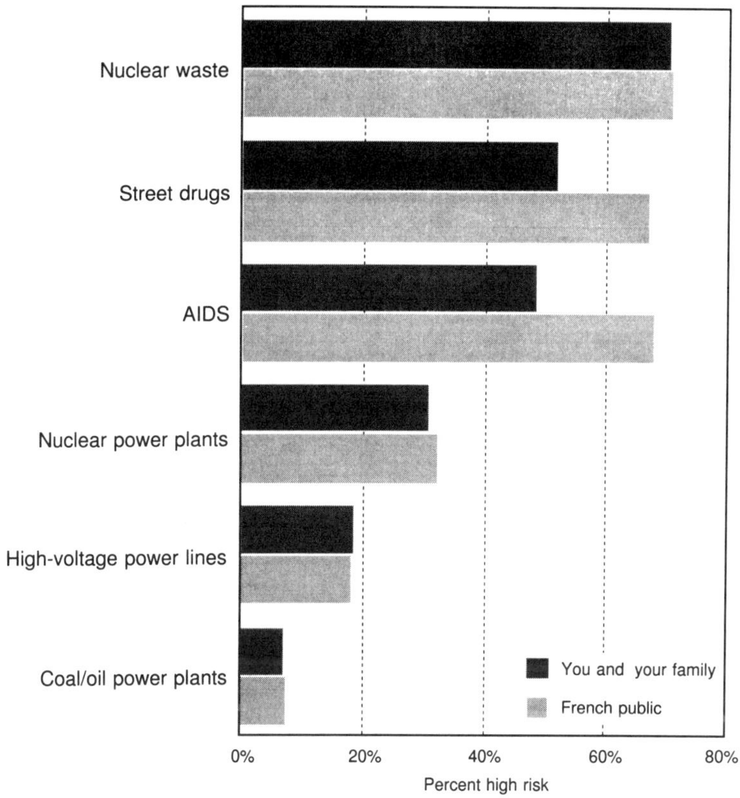

RISK TO FRENCH RESPONDENTS AND THEIR FAMILIES

Figure 7 compares the percentage of high-risk responses when respondents were considering the health risk "to you and your family" with the percentage for the French public as a whole for six items. Nuclear waste, street drugs, and AIDS received more high-risk evaluations than the other items when the reference was to personal or familial risk. There were more high-risk judgments in reference to societal risks than personal risks for street drugs and AIDS. Virtually no differences were seen between personal and societal risks for the other four items.

Figure 8. **PERCEIVED HEALTH RISKS TO RENCH PUBLIC BY GENDER: DIFFERENCE BETWEEN MALES AND FEMALES**

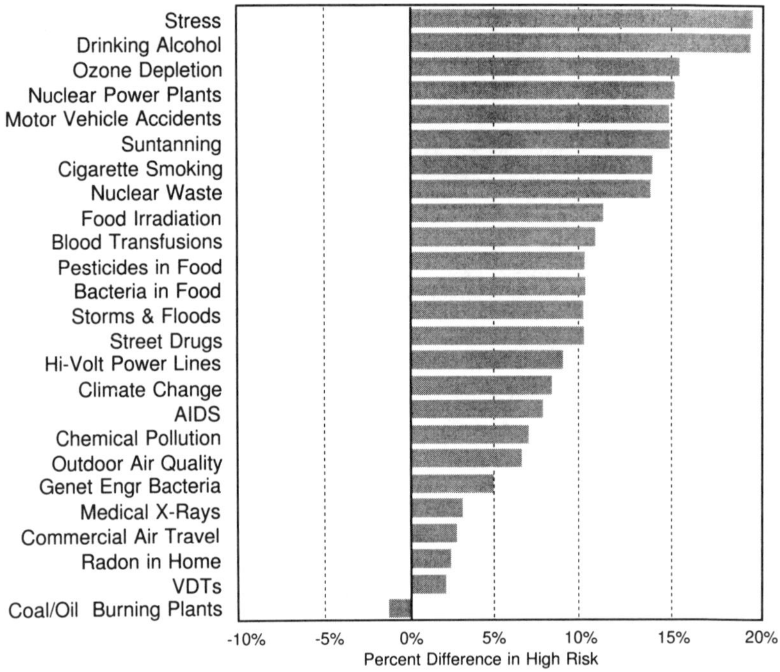

Note: Difference is percent female high-risk responses minus percent male high-risk responses.

GENDER DIFFERENCES: FRENCH SAMPLE

Perceived risk to the French public as a whole was examined for gender differences. Sizable differences were observed between men and women respondents. As can be seen in Figure 8, women were more likely to rate the risk as high for all but one item - coal and oil burning power plants. The greatest differences were seen for stress and drinking alcohol, 19.7% and 19.5%, respectively. Two other items exhibiting more than a 15% difference were ozone depletion and nuclear power plants.

Figure 9. **PERCEIVED HEALTH RISKS TO FRENCH PUBLIC BY AGE: DIFFERENCE BETWEEN 55+ AND UNDER 30 AGE GROUPS**

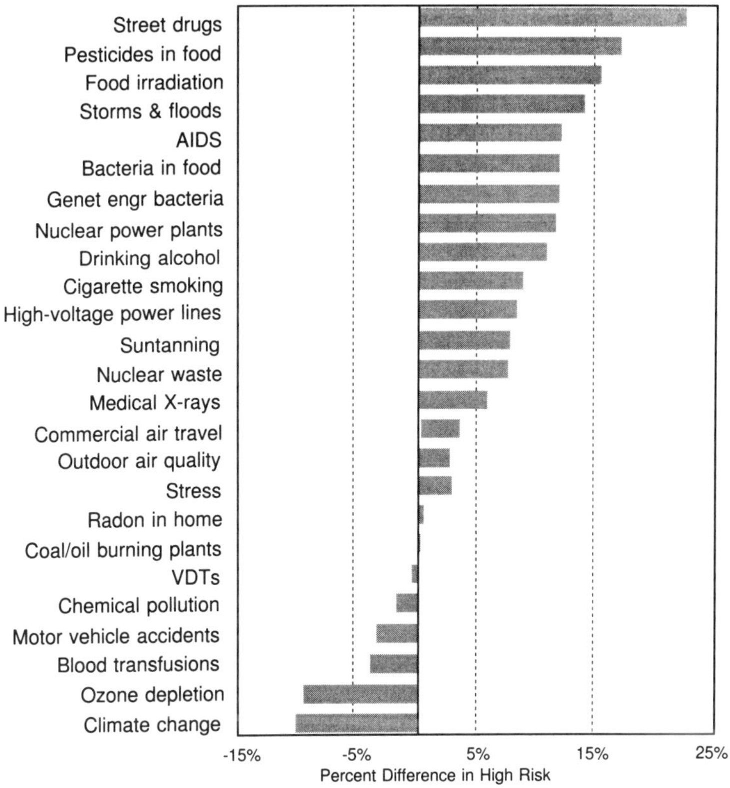

Note: Percent difference is percent 55+ high-risk responses minus under 30 high-risk responses.

AGE DIFFERENCES: FRENCH SAMPLE

Figure 9 portrays the difference in high-risk responses between respondents age 30 or less and age 55 or more. As in the U.S., older respondents were more likely to rate a risk as high. This tendency was particularly evident for street drugs, pesticides in food, food irradiation, storms and floods, AIDS, bacteria in food, genetically engineered bacteria, nuclear power plants (opposite the age trend in the U.S.), and drinking alcohol. Younger persons tended to be somewhat more concerned about climate change and ozone depletion. Very little differences were found for radon in the home, coal and oil burning power plants, and video display terminals.

Figure 10. **PERCEIVED HEALTH RISKS TO FRENCH PUBLIC BY EDUCATION: DIFFERENCE BETWEEN SCHOOL AND COLLEGE-EDUCATED RESPONDENTS**

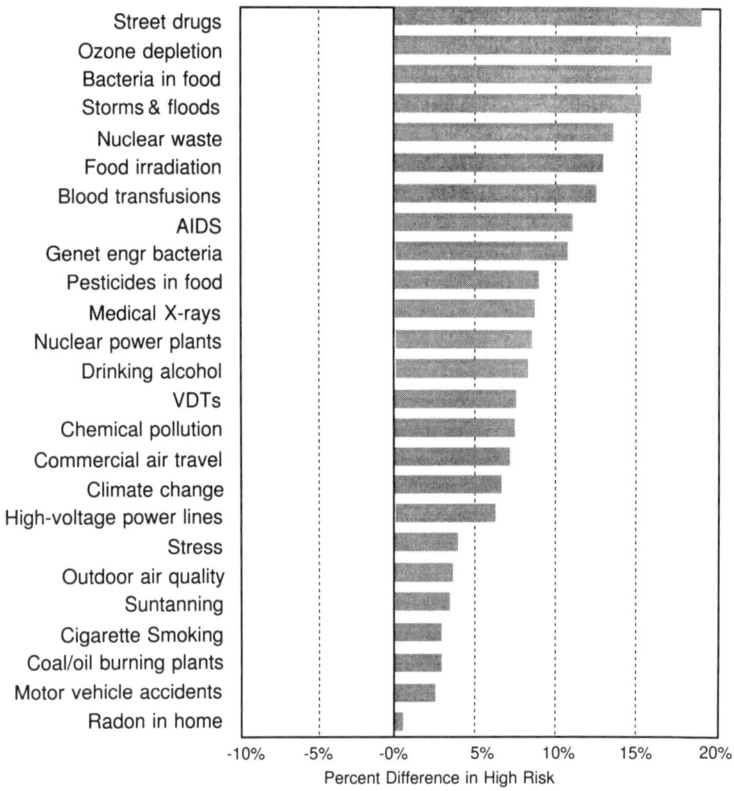

Note: Percent difference is percent high school educated high-risk responses minus percent college-educated high-risk responses.

EDUCATION DIFFERENCES: FRENCH SAMPLE

College-educated respondents were consistently less likely than respondents with high school or less education to rate a risk as high (see Figure 10). People with less formal education were relatively more concerned about street drugs, ozone depletion, bacteria in food, storms and floods, nuclear waste, food irradiation, blood transfusions, AIDS, and genetically engineered bacteria. The maximum difference in the high-risk response associated with education was 18.6% for street drugs.

U.S.–FRANCE COMPARISON: RISK TO THE PUBLIC

Both the French and American respondents recorded more than 40% high-risk responses for eight items: nuclear waste, AIDS, street drugs, cigarette smoking, chemical pollution, ozone

Figure 11. **PERCEIVED HEALTH RISKS TO PUBLIC: U. S. AND FRANCE**

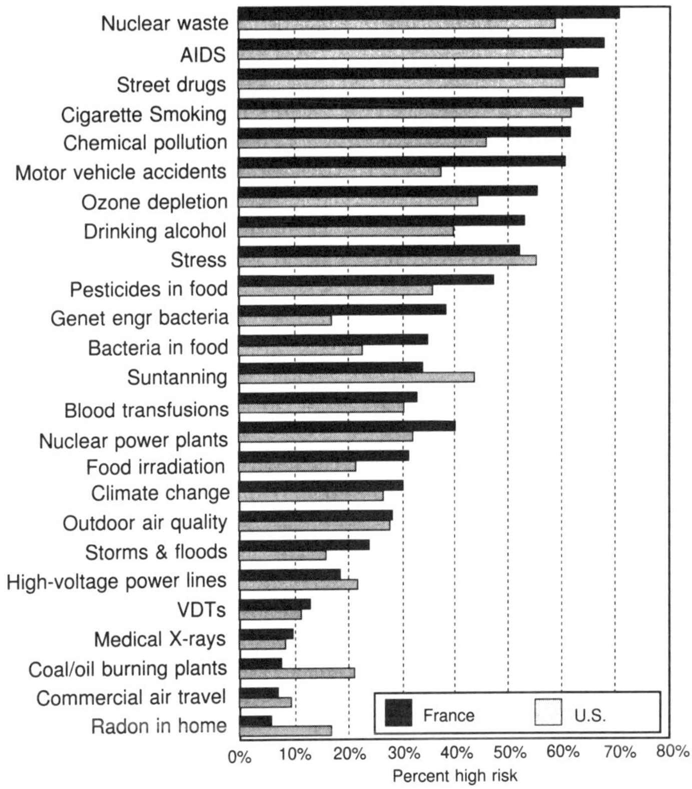

depletion, drinking alcohol, and stress (see Figure 11). Two additional items received more than 40% high-risk responses in France: motor vehicle accidents and pesticides in food. One additional item in the U.S. received over 40% high-risk responses: suntanning.

Overall, French respondents tended to rate the items as higher in risk than did the American respondents. The French respondents produced over 20% more high-risk responses than Americans for two items: motor vehicle accidents and genetically engineered bacteria (see Figure 12). In addition, there were seven items for which the French recorded 10% more high-risk responses than Americans: chemical pollution, drinking alcohol, bacteria in food, nuclear waste, pesticides in food, ozone depletion, and food irradiation. There were only two items for which U.S. respondents recorded more than 10% more high-risk responses than the French, coal and oil burning power plants and radon in the home. Of particular interest is the finding that the smallest difference between the two countries was for nuclear power plants.

Figure 12. **PERCEIVED HEALTH RISKS TO THE PUBLIC: PRECENT DIFFERENCE BETWEEN U.S. AND FRANCE**

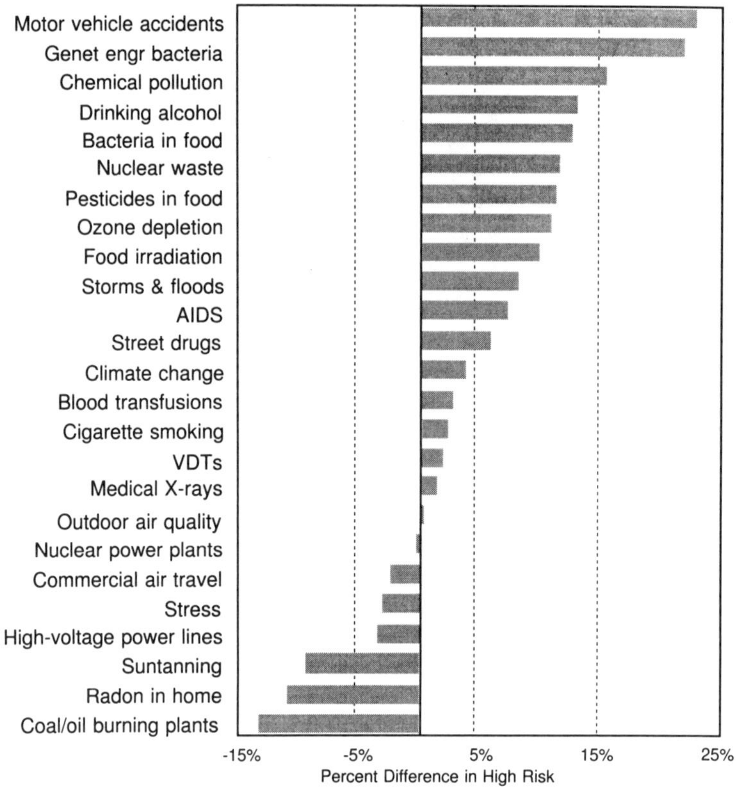

Note: Difference is percent French high-risk responses minus percent U.S. high-risk responses.

U.S.–FRANCE COMPARISON: RISK TO RESPONDENTS AND THEIR FAMILIES

For both U.S. and French respondents, there tended to be somewhat more high-risk judgments in reference to societal risk than personal risk. However, the differences between the two types of risk perception were small except for AIDS and street drugs. For these two items, the magnitude of the differences was much larger in the U.S. than France; for the French the differences between societal and personal risk for AIDS and street drugs was 10.3% and 15.2%, respectively, compared to 33.6% and 35.0% in the U.S. The percentage of high-risk ratings for nuclear power in reference to self and family was about equal in both countries (28.8% in the U.S. vs. 30.8% in France). For nuclear waste, the percentage of high-risk responses to self and family was greater in France (70.8% vs. 55.1% in the U.S.).

U.S.–FRANCE COMPARISON: GENDER ANALYSES

Women in both samples gave more high-risk judgments than men. In the U.S. sample, women were higher on all 25 items compared to 24 in France. In France, the only item for which men recorded more high-risk responses than women was coal and oil burning power plants. In particular, women in both countries rated stress as a much greater risk than did men.

3.2 Word Associations

AFFECTIVE RATINGS

The American respondents provided a total of 3546 associations to the stimulus "nuclear power." The average number of associations per respondent was 2.3. When asked to rate their associations, 46.7% of the respondents gave a positive or very positive rating; 40.6% were negative or very negative, and 12.1% were rated as neutral. There tended to be more extremely negative ratings than extremely positive ones (17.1% versus 10.1%).

French respondents gave a total of 3768 associations to the stimulus "nuclear power" for an average of 2.4 per respondent. The distribution of ratings assigned to these associations by the French respondents was remarkably similar to that of the Americans. Nearly half (49.3%) of the associations were rated as positive or very positive; 10.4% were rated as neutral, and 39.1% as negative or very negative. There were more extreme responses on the negative side than on the positive (19.4% very negative vs. 11.5% very positive).

Within the U.S. sample, men gave more positive ratings to their associations than women (52.5% vs. 40.9%) and people under 30 years of age gave fewer positive ratings than people over 55 (39.2% vs. 60.4%). Similar patterns were found in the French data. French men gave more positive ratings (51.6% vs. 47.3% for women) and younger people gave fewer positive ratings than older persons (43.9% vs. 60.8%).

The affective ratings that respondents made for every image they produced were found to be highly correlated with many of the other responses in the survey, including attitudes of support or opposition to nuclear power. The relationships between affect ratings and other survey variables are presented elsewhere (see Peters & Slovic, 1996).

CONTENT ANALYSIS

The images were categorized into 11 content categories as shown in Tables 2 and 3. There was one large category containing negative associations.[1] For the American sample (Table 2), this category accounted for 41.1% of the associations and included such responses as bombs, death, killing, nuclear waste, accidents, dangerous, risky, hazardous, toxic, and so forth. About one-quarter (26.5%) of the American associations fell into a generally positive category. The positive content category included such associations as electricity, energy, light, economical, good, clean, efficient, necessary, and so forth. In addition, one-third of the images fell into categories that

1 Labeling of these associations as negative in content was done by the research team and should not be confused with the negative affective ratings given by the respondents themselves.

Table 2. **NUCLEAR POWER IMAGES: U.S. DATA**

Associations	Categories	Frequency	Percent
Negative concepts and consequences	A-Bombs/bombs/atomic weapons	289	8.2
	Dangerous/toxic	208	5.9
	War/annihilation/end of world/holocaust	149	4.2
	Waste/garbage/dumps/nuclear waste	142	4.0
	Destruction	128	3.6
	Death/sickness/killing	101	2.9
	Accident/meltdowns/disaster	98	2.8
	Scary/fear	73	2.1
	Bad/negative/dislike/tragedy	60	1.7
	Environmental damage/pollution/unclean	55	1.6
	Unnecessary/opposed/no	36	1.0
	Cost/expensive	32	0.9
	Risk/risky/unsafe	20	0.6
	Contamination	22	0.6
	Leakage	13	0.4
	Other-Negative	22	0.6
	Not near me	8	0.2
	Subtotal	1456	41.1
Positive concepts and consequences	Energy/power/electricity/light	488	13.8
	Clean/clean air/clean energy/cleaner	157	4.4
	Cheap/cost effective/savings	116	3.3
	Positive/unconcerned/beneficial	67	1.9
	Best type energy/conserves resources	34	1.0
	Necessary/need/want more	20	0.6
	Feasible/acceptable	18	0.5
	Economics/money/income	18	0.5
	Employment	10	0.3
	Inexhaustible/endless	6	0.2
	Peace	5	0.1
	Subtotal	939	26.5
Locations	Facilities & their construction	235	6.6
	Locations/specific sites	210	5.9
	Subtotal	445	12.5
Miscellaneous	Future/long lasting	42	1.2
	Alternative/alternate energy source	37	1.0
	Atoms/atomic	21	0.6
	Miscellaneous other	87	2.5
	Subtotal	187	5.3
Radiation/physical states	Radiation, nuclear, fire, hot, chemical processes and states	177	5.0
Concerns	Questions, contested, uncertain, problems, controversy, distrust	87	2.5
Safety/security	Safe, secure, cleanup, control	82	2.3
Ecology	Natural environment, sun, air, food/water supply	53	1.5
Science/technology	Technology, medical research	53	1.5
Descriptive	Incredible, powerful, steam	39	1.1
Information/Knowledge	Media, curiosity, knowledge	28	0.8
	Total Number of Associations	3546	

Table 3. **NUCLEAR POWER IMAGES: FRENCH DATA**

Associations	Categories	Frequency	Percent
Negative concepts and consequences	A-Bombs/bombs/atomic weapons	374	9.9
	Dangerous/toxic	251	6.7
	War/annihilation/end of the world/holocaust	181	4.8
	Environmental damage/pollution/unclean	133	3.5
	Accident/meltdowns/disaster	107	2.8
	Scary/fear	86	2.3
	Death/sickness/killing	57	1.5
	Waste/garbage/dumps/nuclear waste	46	1.2
	Destruction	42	1.1
	Risk/risky/unsafe	39	1.0
	Unnecessary/opposed/no	8	0.2
	Contamination	7	0.2
	Leakage	6	0.2
	Subtotal	1337	35.5
Positive concepts and consequences	Energy/power/electricity/light	713	18.9
	Positive/unconcerned/beneficial	92	2.4
	Economics/Money/Income	32	0.9
	Necessary/need/want more	24	0.6
	Clean/clean air/clean energy/cleaner	12	0.3
	Employment	8	0.2
	Subtotal	881	23.4
Locations	Facilities & their construction	452	12.0
	Locations/specific sites	235	6.2
	Subtotal	687	18.2
Miscellaneous	Atoms/atomic	105	2.8
	Future/long lasting	69	1.8
	Political processes	59	1.6
	Miscellaneous other	50	1.3
	Subtotal	283	7.5
Radiation/physical states	Radiation, nuclear, fire, hot, chemical processes and states	130	3.5
Safety/security	Safe, secure, cleanup, control	52	1.4
Ecology	Natural environment, sun, air, food/water, supply	133	3.5
Science/technology	Science, medical research	219	5.8
Information/knowledge	Information/knowledge - other	15	0.4
Uncategorized	Uncategorized	31	0.8
	Total Number of Associations	3768	

were neither clearly negative nor positive by content, such as specific locations or facilities (e.g., TMI, Hanford), components of nuclear plants (e.g., tower, dome), or physical states (e.g., radiation, heat, fission).

The frequency distribution of content categories in the French data is shown in Table 3. About one-third (35.5%) of the French associations fell in the negative category. Positive associations accounted for 23.4% of the responses.

The historical link between nuclear power and nuclear weapons was noted by Smith (1988), who observed, "nuclear energy was conceived in secrecy, born in war, and first revealed to the world in horror. No matter how much proponents try to separate the peaceful from the weapons atom, the connection is firmly embedded in the minds of the public" (p. 1606). In light of this link, it is of interest to examine the degree to which the word associations reflect that link. The evidence is clear. The category labeled "A-bombs, bombs, atomic weapons" was the second largest in both countries, behind only the category "Energy, power, electricity, light." The A-bomb category, combined with the category "War, annihilation, end of world, holocaust," totaled 12.4% of the U.S. associations and 14.7% of the French associations. Frequent associations to destruction and death may also be related to the weapons connection.

Comparing the U.S. and French content distributions shows a basically similar pattern, except that negative associations were slightly more frequent in the U.S. sample. There were also more ecological references in the French data (3.5%) than in the American data (1.5%). The same held true for science and technology associations (5.8% in France vs. 1.5% in the U.S.).

An earlier survey by Slovic, Flynn, and Layman (1991) asked a representative sample of Americans to associate to the stimulus prompt "underground nuclear waste repository." Negative concepts and consequences clearly dominated the nuclear waste imagery and accounted for 56.2% of the responses. This was considerably higher than the 41.1% negative content to the stimulus "nuclear power" in the present U.S. survey.

In the U.S. survey, women were somewhat more likely than men to give images that fell in the clearly negative content categories (44.9% vs. 37.2%). No such differences were found in France. In the U.S., older respondents (age 55 or more) were more likely to provide positive images (34.3%) compared to respondents under age 30 (20.8% positive). A similar, but weaker pattern was found in France, where 27.3% of those age 55 or older produced positive images compared to 23.2% of those age 30 or younger.

3.3 Support for Nuclear Power

U.S.: NUCLEAR POWER

A number of statements were posed to assess respondents' support for nuclear power and attitudes about various aspects of nuclear power. U.S. respondents were divided in their support. Even if their community was faced with a potential shortage of electricity, only 46.7% agreed that a new nuclear power plant should be built to supply that electricity (see Table 4).[2] The level of agreement was similar (45.6%) when asked, "In light of health concerns about acid rain, damage to the ozone layer, and climate change associated with the burning of coal and oil, America should rely more heavily on nuclear power to meet its future electricity needs." Again,

[2] Table 4 illustrates a characteristic difference between French and American respondents. The French more frequently used the extreme response categories, "strongly disagree" and "strongly agree". This occurred with almost every attitudinal item. This may be due to a response bias, rather than to any greater bipolarity in French attitudes. On the assumption that this is a response bias, we shall combine "strongly disagree" with "disagree" responses and "strongly agree" with "agree" responses when refering to the results in the text.

Table 4: **SUPPORT FOR NUCLEAR POWER**

			Strongly disagree	Disagree	Agree	Strongly agree	Don't know/ no answer
Q5	If your community was faced with a potential shortage of electricity, do you strongly agree . . . that a new nuclear power plant should be built to supply that electricity	U.S.	18.7%	31.5%	38.8%	7.9%	3.1%
		France	29.7	20.7	35.5	13.2	.8
Q65	In light of health concerns about acid rain, damage to the ozone layer, and climate change associated with the burning of coal and oil, America [France] should rely more heavily upon nuclear power to meet its future electricity needs	U.S.	6.2	43.5	42.5	3.1	4.8
		France	19.5	28.0	31.2	19.9	1.5
Q68	In order to avoid importing energy from other countries to meet our future electricity needs America [France] should rely more heavily upon nuclear power	U.S.	4.4	46.0	43.4	3.2	3.0
		France	22.3	33.3	26.2	16.8	1.5
Q77	The nuclear power industry says that it is now possible to build a new generation of nuclear power plants that will be safer than existing plants. Assuming the nuclear power industry is correct, I would support such a new generation of nuclear plants to supply the country's future electricity needs	U.S.	4.3	27.6	59.5	6.3	2.2
		France	10.5	12.6	40.3	35.9	.6
			Not acceptable	Slightly acceptable	Moderately acceptable	Very acceptable	Don't know/ no answer
Q104	Nuclear power	U.S.	25.8	20.6	34.8	18.1	.7
		France	21.9	14.4	44.5	19.1	.1

Note: The four response categories to these five items were coded 1 (strongly disagree; not acceptable) to 4 (strongly agree; very acceptable) and averaged for each respondent. The mean response was labeled the nuclear support index. The reliability of this index, as measured by Cronbach's alpha, was .84.

fewer than half (46.6%) of the respondents agreed that America should rely more heavily on nuclear power to avoid importing energy from other countries to meet our future electricity needs. However, nearly two-thirds of the U.S. respondents agreed with the statement "The nuclear power industry says that it is now possible to build a new generation of nuclear power plants that will be safer than existing plants. Assuming the nuclear power industry is correct, I would support such new generation nuclear power plants to supply the country's future electricity needs."

Table 5: **ACCEPTABILITY OF ALTERNATIVE WAYS TO PRODUCE ELECTRICITY**

		Not acceptable	Slightly acceptable	Moderately acceptable	Very acceptable	Don't know/ no answer
Q101 Oil	U.S.	10.3%	21.5%	47.8%	19.8%	.5%
	France	4.1	8.1	57.8	29.2	.7
Q102 Solar power	U.S.	1.5	5.1	20.7	71.8	.9
	France	1.5	2.6	13.4	82.3	.1
Q103 Natural gas	U.S.	3.7	9.2	43.3	43.5	.3
	France	1.9	3.1	39.7	54.9	.3
Q104 Nuclear power	U.S.	25.8	20.6	34.8	18.1	.7
	France	21.9	14.4	44.5	19.1	.1
Q105 Hydroelectric power	U.S.	3.1	7.3	28.8	59.7	1.1
	France	1.6	2.7	23.0	72.5	.1
Q106 Wind power	U.S.	3.4	8.2	19.9	68.0	.5
	France	3.8	4.6	15.9	75.5	.3
Q107 Coal	U.S.	18.4	25.5	43.2	12.4	.5
	France	5.5	9.1	53.3	32.0	.1

Respondents were presented with seven ways to produce electricity and asked how acceptable each of these ways is to them for meeting the nation's future energy needs. An overwhelming majority (92.5%) considered solar power moderately or very acceptable (see Table 5). Solar power was followed by hydroelectric power, wind power, and natural gas, with 86.8 to 88.5% of the respondents indicating these were acceptable ways to produce electricity. Nuclear power was the least acceptable method, with only 52.9% moderately acceptable or very acceptable responses.

A majority of the U.S. respondents (80.5%) thought other options for providing electricity should be developed instead of building more nuclear power and fossil fuel plants (see Q80 in Table 6). More than two-thirds of the respondents agreed that "Having a nuclear power plant nearby makes other people think that a community is a less attractive place" (Q86). Two-thirds of the respondents felt that nuclear power was a technological achievement of which our nation can be proud (Q69). However, respondents were divided over whether nuclear power is essential to our nation's security (Q73) or economic well-being (Q76), with 51.4% and 53.1% agreement, respectively. Only one-third of the respondents thought the nuclear industry is capable of managing its wastes safely (Q78).

Respondents were divided over whether most scientists agree that the risks of nuclear power are acceptable (55.6% agreement in Q84). Similarly, only 52.1% agreed that differences of opinion about the risks of nuclear power can be resolved by scientific data and analysis (Q90). About

Table 6: **CHARACTERISTICS OF NUCLEAR POWER**

		Strongly disagree	Disagree	Agree	Strongly agree	Don't know/ no answer
Q69 Nuclear power is a technological achievement of which our nation can be proud	U.S.	2.8%	28.5%	60,8%	4.6%	3.2%
	France	11.6	17.1	36.7	33.9	.7
Q71 We should stop using nuclear power plants because we do not know how to store radioactive wastes safely	U.S.	2.1	39.4	44.4	10.6	3.5
	France	12.6	27.3	26.2	33.1	.8
Q73 Nuclear power is essential to our nation's security	U.S.	4.0	40.5	48.0	3.4	4.2
	France	20.6	28.5	26.8	22.1	2.1
Q76 Nuclear power is essential to our nation's economic well-being	U.S.	3.2	39.2	50.3	2.8	4.4
	France	9.9	22.3	38.3	28.0	1.5
Q78 The nuclear industry is capable of managing its wastes safely	U.S.	9.9	54.5	31.1	1.5	3.0
	France	32.6	39.6	16.5	10.1	1.2
Q80 We should develop other options for providing electricity instead of building more nuclear power and fossil fuel plants	U.S.	.4	17.4	63.4	17.1	1.7
	France	4.1	9.4	31.7	53.9	.9
Q82 The production of nuclear power contributes to the production of nuclear weapons	U.S.	2.4	46.0	44.2	2.8	4.6
	France	22.3	21.7	23.2	27.3	5.4
Q84 Most scientists agree that the risks of nuclear power are acceptable	U.S.	2.4	33.7	54.0	1.6	8.2
	France	18.2	25.7	34.6	19.6	1.8
Q85 Nuclear power is immoral because it imposes risks upon future generations without their consent	U.S.	3.1	42.6	45.3	6.3	2.6
	France	14.5	24.6	23.5	36.2	1.2
Q86 Having a nuclear power plant nearby makes other people think that a community is a less attractive place	U.S.	1.3	27.8	62.9	6.2	1.9
	France	9.8	16.1	27.1	46.5	.5
Q90 Differences of opinion about the risks of nuclear power can be resolved by scientific data and analysis	U.S.	2.8	42.2	49.9	2.2	2.8
	France	10.5	23.4	40.2	22.7	3.2

half (51.6%) thought that "Nuclear power is immoral because it imposes risks on future generations without their consent" (Q85).

Eight items were included to compare nuclear power risks to those of coal and oil on four characteristics: whether the risks were well understood or not well understood by science, well understood or not well understood by the public, controllable or not controllable by science, and catastrophic or not catastrophic. Table 7 displays the results. Nuclear power risks were perceived as less well understood by both science and the public than coal and oil risks, less controllable, and more catastrophic than coal and oil risks.

Table 7: **CHARACTERISTICS OF NUCLEAR POWER AND COAL AND OIL RISKS**

		Nuclear power	Coal and oil
Well understood by science	U.S.	55.1%	81.9%
	France	60.6	80.0
Well understood by public	U.S.	8.1	39.4
	France	13.9	49.0
Controllable by science	U.S.	53.0	67.8
	France	50.3	78.8
Catastrophic	U.S.	68.8	37.5
	France	80.3	45.4

GENDER DIFFERENCES: U.S. SAMPLE

Men tended to be more supportive of nuclear power than women. Men recorded 12–13% more agree responses in support of nuclear power if faced with electricity shortages, to avoid importing energy from other countries, and in light of health concerns about acid rain, damage to the ozone layer, and climate change. Men were also more likely to support a new generation of nuclear power plants, with 8.4% more agree responses. Women were more likely to believe we should stop using nuclear power plants because we do not know how to store radioactive wastes safely, with 11.7% more agree responses than men. Men were more likely to indicate that nuclear power was an acceptable method of producing electricity than women, with 13.5% more acceptable responses.

Women were more likely to indicate that nuclear power was immoral because of the risks it imposes on future generations, with 13.4% more agreement with this statement than men. They also were more likely than men to agree that a nuclear power plant makes the nearby area unattractive. Men were more likely to indicate that most scientists agree that the risks of nuclear power are acceptable and that nuclear power is a technological achievement of which our nation can be proud. In addition, men were more likely than women to believe that nuclear power was essential to our national economy and security.

In addition, men were more likely to believe nuclear power risks are well understood and controllable by science than women. Women were more likely to characterize nuclear power risks as catastrophic. Gender differences in the characterization of risks from coal and oil were small.

AGE DIFFERENCES: U.S. SAMPLE

There appeared to be little difference between older and younger respondents in terms of support for nuclear power. Respondents age 55 or more were somewhat more likely to agree with the statement that to avoid importing energy from other countries Americans should rely more on nuclear power. Older respondents (age 55 and older) tended to indicate that oil and nuclear power were more acceptable ways of producing electricity than younger respondents (age 30 or younger), with about eight percent more acceptable responses.

Older respondents were more likely than younger respondents to indicate that nuclear power is essential to our national economy and to our national security and that nuclear power is a technological achievement of which to be proud, with 12–14% more agreement. They also were somewhat more likely to think the nuclear industry is capable of managing nuclear wastes safely. Younger respondents were somewhat more likely to believe nuclear power is immoral because it imposes risks on future generations and that a nuclear power plant nearby makes a place unattractive. Younger respondents were more likely to indicate that nuclear power and coal and oil risks are catastrophic.

EDUCATION DIFFERENCES: U.S. SAMPLE

There were no large educational differences on the nuclear power support questions; high school educated respondents were somewhat more likely than the college educated respondents to indicate that coal and oil were acceptable ways to produce electricity. College-educated respondents were somewhat more likely to indicate that nuclear power was an acceptable method of producing electricity. They also were more likely to feel nuclear power risks were controllable by science.

Those with a high school education produced 16.1% more agree responses than those with a college education when asked whether nuclear power was immoral because it imposes risks on future generations. They also were more likely to agree that nuclear power contributes to the production of nuclear weapons, with 12% more agreement than college educated respondents.

FRANCE: NUCLEAR POWER

Generally, the French were divided in their support for nuclear power. About 49% expressed support for new nuclear power plants if their community was faced with a potential shortage of electricity (see Table 4). Similarly, 51.1% supported nuclear power in light of health concerns about acid rain, damage to the ozone layer, and climate change associated with burning of coal and oil. Only 43% wanted to rely more heavily on nuclear power to avoid importing energy from other countries. However, three-quarters of the respondents would support a new generation of nuclear plants if they could be assured that the plants were safer than existing plants. Nearly 60% of the French respondents agreed that "We should stop using nuclear power plants because we do not know how to store radioactive wastes safely."

An overwhelming majority of the French respondents indicated that solar power, hydroelectric power, natural gas, and wind power would be moderately acceptable or very acceptable ways to produce electricity (see Table 5). Nuclear power was found to be the least acceptable method; however, 63.6% still found it a moderately or very acceptable way to produce electricity.

Most of the respondents (85.6%) agreed that "We should develop other options for providing electricity instead of building more nuclear power and fossil fuel plants (see Table 6). While nearly three-quarters indicated that having a nuclear power plant nearby "makes other people think that a community is a less attractive place," a majority (70.6%) also thought nuclear power was a "technological achievement of which our nation can be proud." Two-thirds of the respondents thought nuclear power is essential to the national economic well-being; however, only 48.9% thought it was essential to national security. Respondents were divided about whether the production of nuclear power contributes to the production of nuclear weapons (50.5% agreement; 44.0% disagreement).

Nearly 60% of the French sample agreed that nuclear power was immoral because it imposed risks on future generations without their consent. Just over half (54.2%) thought that most scientists agree that the risks of nuclear power are acceptable. Even more (62.9%) thought that scientific data and analysis could resolve differences of opinion about the risks of nuclear power. However, only 26.6% agreed that the nuclear industry was capable of managing its wastes safely.

In addition, the French thought nuclear power risks were less well understood by both science and the public, less controllable, and more catastrophic than coal and oil risks (see Table 7).

GENDER DIFFERENCES: FRENCH SAMPLE

French men were more likely to be supportive of nuclear power than women. Twelve percent more men than women expressed support for new nuclear plants if their community was faced with electricity shortages. Men also expressed somewhat more support for nuclear power in light of acid rain, damage to the ozone layer, climate change, and to avoid importing energy. Women were more likely to agree that we should stop using nuclear power plants because we don't know how to store wastes safely (14.6% more agreement than men). Male support for nuclear power was also seen in the responses to the acceptability of various forms of producing electricity. Men recorded 12.8% more moderately acceptable or very acceptable responses than women with respect to nuclear power. There were no other sizable gender differences in regard to the acceptability of other forms of producing electricity.

Women were more likely to indicate that nuclear power is immoral because of the risks it imposes on future generations; women produced 14.1% more agree responses to this statement. They also were more likely to believe that a nuclear power plant makes a community less attractive. Men were more likely to think that nuclear power is a "technological achievement of which our nation can be proud," with 15.7% more agree responses than women. Men also were more likely to think that nuclear power is essential to the national economy and to national security and to indicate that most scientists agree that the risks of nuclear power are acceptable. Men also tended to be more likely to believe that nuclear power risks were well understood and controllable by science.

AGE DIFFERENCES: FRENCH SAMPLE

Age differences in support for nuclear power were rather small. Respondents age 55 or over were somewhat more likely to indicate that nuclear power was an acceptable way of producing electricity, with 8.9% more acceptable responses than respondents under age 30. Older respondents were also more likely to agree that nuclear power was essential to the national security with 9.3% more agreement than younger respondents. Older respondents tended to be more likely than younger respondents to think the nuclear power industry could manage radioactive wastes safely.

There were few age differences in how characteristics of nuclear power risks were seen. However, younger respondents were much more likely to see coal and oil risks as catastrophic and well understood by science.

EDUCATION DIFFERENCES: FRENCH SAMPLE

There was a greater tendency for those with a high school education or less to think that we should stop using nuclear power because we do not know how to store the wastes safely. They also were slightly more likely to believe we should rely more on nuclear power in light of acid rain, ozone depletion, and climate change. No other educational differences in support for nuclear power were found. Educational background did not appear to have much influence in attitudes regarding the acceptability of various forms of producing electricity.

Respondents with a high school education were more likely than those with a college education to indicate that nuclear power was immoral because it imposes risks on future generations without their consent, with 12.8% more agree responses. College educated respondents were somewhat more likely to believe that scientists agree that the risks of nuclear power are acceptable. Those with higher education levels were more likely to think nuclear power risks and coal and oil burning risks were well understood by science.

U.S.–FRANCE COMPARISON: SUPPORT FOR NUCLEAR POWER

French respondents were slightly more likely to express support for nuclear power than U.S. respondents (see Tables 4 and 8). In particular, French respondents gave 10.4% more agree responses than Americans when asked if they would support a new generation of nuclear power plants if they could be assured that the plants would be safer than the existing plants.

Table 8: **NUCLEAR POWER SUPPORT: U.S.–FRANCE COMPARISON**

		Difference
Q104	Nuclear power	10.7%
Q77	The nuclear power industry says that it Is now possible to build a new generation of nuclear power plants that will be safer than existing plants. Assuming the nuclear power industry is correct, I would support such a new generation of nuclear power plants to supply the country's future electricity needs	10.39
Q65	In light of health concerns about acid rain, damage to the ozone layer, and climate change associated with the burning of coal and oil, Americans [French] should rely more heavily upon nuclear power to meet its future electricity needs	5.40
Q5	If your community was faced with a potential shortage of electricity, do you strongly agree . . . that a new nuclear power plant should be built to supply that electricity	2.08
Q68	In order to avoid importing energy from other countries to meet our future electricity needs America [France] should rely more heavily upon nuclear power	-3.59

Note: Values shown are differences in percent agree responses for France minus percent agree responses in the U.S. except for Q104 where the difference is in terms of percent saying "moderately acceptable" or "very acceptable".

French respondents were much more likely to consider coal an acceptable way to produce electricity, with 29.7% more acceptable responses than American respondents (see Table 9). The French were also more likely than Americans to find oil and nuclear power moderately or very acceptable.

Table 9: **ACCEPTABILITY OF ALTERNATIVE WAYS TO PRODUCE ELECTRICITY: U.S.–FRANCE COMPARISON**

		Difference
Q107	Coal	29.7%
Q101	Oil	19.4
Q104	Nuclear power	10.7
Q103	Natural gas	7.8
Q105	Hydroelectric power	7.0
Q106	Wind power	3.5
Q102	Solar power	3.2

Note: Values shown are differences in percent agree responses for France minus percent agree responses in the U.S. except for Q104 where the difference is in terms of percent saying "moderately acceptable" or "very acceptable."

Table 10: **CHARACTERISTICS OF NUCLEAR POWER: U.S.–FRANCE COMPARISON**

		Difference
Q76	Nuclear power is essential to our nation's economic well-being	13.21%
Q90	Differences of opinion about the risks of nuclear power can be resolved by scientific data and analysis	10.79
Q85	Nuclear power is immoral because it imposes risks upon future generations without their consent	8.09
Q69	Nuclear power is a technological achievement of which our nation can be proud	5.17
Q80	We should develop other options for providing electricity instead of building more nuclear power and fossil fuel plants	5.06
Q86	Having a nuclear power plant nearby makes other people think that a community is a less attractive place	4.50
Q71	We should stop using nuclear power plants because we do not know how to store radioactive wastes safely	4.26
Q82	The production of nuclear power contributes to the production of nuclear weapons	3.49
Q84	Most scientists agree that the risks of nuclear power are acceptable	-1.36
Q73	Nuclear power is essential to our nation's security	-2.55
Q78	The nuclear industry is capable of managing its wastes safely	-9.96

Note. Values shown are differences in percent agree responses for France minus percent agree responses in the U.S.

When asked if nuclear power was essential to their national economy, French respondents recorded 13.2% more agree responses than U.S. respondents (see Table 10). They were also more likely than Americans to believe that science can resolve differences of opinions about nuclear power risks and that nuclear power is immoral because it imposes risks on future generations.

3.4 Trust in Experts, Government, and Science

U.S.: TRUST IN EXPERTS, GOVERNMENT, AND SCIENCE

Trust in the government, experts, and science was the focus of seven questions (see Table 11). Disagreement that government and industry can be trusted to make proper decisions to manage technological risks was indicated by 64.7% of the U.S. respondents (Q119). Lack of trust was further indicated by three-quarters of the respondents who disagreed that decisions about health risks should be left to the experts (Q64). In addition, a vast majority (84.2%) lacked confidence in the ability of public health officials to take care of really serious health problems (Q55). When risks involved nuclear power, respondents were divided. Just over half (54.3%) of the respondents lacked trust in the experts and engineers who build, operate, and regulate nuclear power plants (Q88). Slightly under half (46.5%) lacked confidence in experts and scientists to make accurate estimates of the risks from nuclear power (Q50). In response to Q90, respondents were also divided about whether differences of opinion about the risks of nuclear power can be resolved by scientific data and analysis (45.0% disagreed; 52.1% agreed). Another question was included to gauge one's trust in the use of animal studies to determine a chemical's risk to humans (Q54). There was a sizable level of trust expressed when asked "If an animal study produces evidence that a chemical causes cancer in animals, then we can be reasonably sure that the substance will cause cancer in humans" (63.2% agreement).

In general, men exhibited somewhat more trust in experts, government, and science than did women. Men were more likely to agree that experts and engineers can be trusted to build, operate, and regulate nuclear power plants and that government and industry can be trusted to make proper decisions about risks from technology. Differences on the other trust items were minor.

FRANCE: TRUST IN EXPERTS, GOVERNMENT, AND SCIENCE

The French respondents exhibited moderately high levels of trust in experts, government, and science. Approximately two-thirds of the respondents agreed that decisions about health risks should be left to the experts and that experts and engineers who build, operate, and regulate nuclear power plants could be trusted (see Table 11). Similarly, 62.9% of the French agreed that differences of opinion about the risks of nuclear power can be resolved by scientific data and analysis. Respondents were more divided about whether experts and/or scientists are able to make accurate estimates of the risks from nuclear power (55.2% agreement). Less confidence was demonstrated about the ability of the government and industry to make proper decisions to manage risks from technology (39.6% agreement). Trust in certain kinds of scientific studies was demonstrated by two-thirds of the respondents who agreed that if an animal study that

Table 11: **TRUST IN EXPERTS, GOVERNMENT, AND SCIENCE**

			Strongly disagree	Disagree	Agree	Strongly agree	Don't know/ no answer
Q50	Experts/scientists are able to make accurate estimates of the risk from nuclear power	U.S.	6.2%	40.3%	46.6%	4.9%	2.0%
		France	17.1	26.9	27.7	27.5	.8
Q54	If a scientific study produces evidence that a substance causes cancer in animals, then we can be reasonably sure that the substance will cause cancer in humans	U.S.	2.0	32.9	56.9	6.3	1.9
		France	10.2	23.3	36.1	28.5	1.9
Q55	When there is a really serious health problem, the public health officials will take care of it. Until they alert me about a specific problem, I don't really have to worry	U.S.	20.8	63.4	14.0	1.4	.5
		France	29.5	28.4	22.8	18.5	.8
Q64	Decisions about health risks should be left to the experts	U.S.	9.8	66.4	21.8	1.5	.6
		France	14.4	17.3	25.9	42.0	.4
Q88	We can trust the experts and engineers who build, operate, and regulate nuclear power plants	U.S.	7.0	47.3	41.4	2.0	2.3
		France	11.4	21.7	39.7	26.7	.5
Q90	Differences of opinion about the risks of nuclear power can be resolved by scientific data and analysis	U.S.	2.8	42.2	49.9	2.2	2.8
		France	10.5	23.4	40.2	22.7	3.2
Q119	Our government and industry can be trusted with making the proper decisions to manage the risks from technology	U.S.	10.3	54.4	32.9	1.1	1.4
		France	21.0	38.2	28.0	11.6	1.2

produces evidence that a substance causes cancer in animals then we can be reasonably sure the substance will cause cancer in humans.

In general, men were somewhat more likely to express trust in experts, government, and science. Men recorded 14.5% more agree responses to the statement "We can trust experts and engineers who build, operate, and regulate nuclear power plants." Men also tended to express more trust in science to resolve differences of opinion about nuclear power risks and that government and industry could be trusted to manage technological risks.

U.S.–FRANCE COMPARISON: TRUST IN EXPERTS, GOVERNMENT, AND SCIENCE

Overall, the French respondents expressed much higher levels of trust in experts, government, and science to manage risks. The French produced 44.7% more agree responses when asked if decisions about health risks should be left to the experts (see Table 12). The French were much more likely to trust health care officials to manage serious health problems, with 26% more agree responses than American respondents. The French respondents also produced 23.0% more agree responses when asked if the experts and engineers could be trusted to manage nuclear power plants.

Table 12: **TRUST IN EXPERTS, GOVERNMENT, AND SCIENCE: U.S.–FRANCE COMPARISON**

		Difference
Q64	Decisions about health risks should be left to the experts	44.72%
Q55	When there is a really serious health problem, the public health officials will take care of it. Until they alert me about a specific problem, I don't really have to worry	25.95
Q88	We can trust the experts and engineers who build, operate, and regulate nuclear power plants	23.00
Q90	Differences of opinion about the risks of nuclear power can be resolved by scientific data and analysis	10.79
Q119	Our government and industry can be trusted with making the proper decisions to manage the risks from technology	5.62
Q50	Experts/scientists are able to make accurate estimates of the risks from nuclear power	3.70
Q54	If a scientific study produces evidence that a substance causes cancer in animals, then we can be reasonably sure that the substance will cause cancer in humans	1.35

Note: Values shown are differences in percent agree responses for France minus percent agree responses in the U.S.

3.5 Worldviews

U.S.: WORLDVIEWS

Response distributions for questions pertaining to worldviews are shown in Table 13. Three questions assessed the fatalism worldview. American respondents tended not to hold fatalistic views. Only one-third of the respondents felt that they had very little control over risks to their health (Q63). Even fewer, 18.3%, agreed with the statement "It's no use worrying about public affairs; I can't do anything about them anyway" (Q122). In addition, a vast majority (83.9%) agreed with the statement in Q48 that "People can offset health risks from pollution by improving their individual lifestyles, such as exercising and eating properly."

Four questions were designed to assess the hierarchy worldview. Little support was indicated for this view by the American respondents. Only 14.1% of the respondents disagreed that "People in positions of authority tend to abuse their power" (Q110). Similar disagreement (14.0%) was found in responses to the statement "Those in power often withhold information about things that are harmful to us" (Q126). Some support (58.9%) was shown for "the need to pull together and support the energy choices our government has made" (Q114).

Considerable support was indicated for the egalitarian worldview. Over three-quarters of the respondents agreed that "If people in this country were treated more equally we would have few problems" (Q113). Two-thirds expressed agreement that "what this world needs is a more equal

Table 13: **WORLDVIEWS**

		Strongly disagree	Disagree	Agree	Strongly agree	Don't know/ no answer
Technological enthusiasm						
Q108 A high-technology society is important for improving our health and social well-being	U.S.	.8 %	12.4 %	58.7 %	27.4 %	.7 %
	France	3.4	7.6	29.0	59.5	.5
Q117 Technological development is destroying nature	U.S.	2.8	33.9	53.7	7.5	2.1
	France	5.5	19.4	31.9	42.8	.4
Conservative/authoritarian						
Q109 I am in favor of capital punishment	U.S.	5.0	20.5	47.8	25.3	1.5
	France	31.8	13.5	18.9	35.0	.8
Q125 The police should have the right to listen to private phone calls to investigate a crime	U.S.	15.5	47.6	31.9	3.7	1.4
	France	34.1	18.5	20.8	26.2	.4
Future generations						
Q118 Our technologies might impose risks upon future generations, but I believe future generations will be able to take care of themselves	U.S.	5.6	43.5	46.2	3.4	1.4
	France	7.4	18.1	36.8	36.4	1.4
Economic growth						
Q115 Continued economic growth can only lead to pollution and depletion of natural resources	U.S.	5.1	49.2	39.0	4.8	2.0
	France	8.5	21.9	27.9	40.5	1.0
Q124 Continued economic growth is necessary to improve our quality of life	U.S.	1.4	17.7	70.7	9.5	.8
	France	6.3	13.2	36.6	43.3	.6
Fatalism/control						
Q 48 People can offset health risks from pollution by improving their individual lifestyles, such as exercising and eating properly	U.S.	1.9	13.6	55.0	28.9	.7
	France	5.8	6.7	31.0	56.5	0.0
Q 63 I feel that I have very little control over risks to my health	U.S.	6.6	58.3	31.0	3.7	.3
	France	7.9	12.5	31.0	48.3	.4
Q122 It's no use worrying about public affairs; I can't do anything about them anyway	U.S.	17.9	63.4	15.7	2.6	.4
	France	42.4	31.4	11.2	14.6	.5
Hierarchy						
Q110 People in positions of authority tend to abuse their power	U.S.	.5	13.6	58.5	26.7	.7
	France	4.0	6.5	24.1	65.4	.1
Q114 We need to pull together and support the energy choices our government has made	U.S.	3.1	36.0	51.5	7.4	2.1
	France	8.6	20.4	27.7	42.1	1.2
Q126 Those in power often withhold information about things that are harmful to us	U.S.	1.0	13.0	65.1	20.0	.9
	France	3.2	6.6	30.0	59.2	1.0

			Strongly disagree	Disagree	Agree	Strongly agree	Don't know/ no answer
Egalitarian view							
Q113	If people in this country were treated equally we would have fewer problems	U.S.	1.3	19.8	58.9	19.7	.3
		France	6.2	11.2	23.4	58.2	1.0
Q121	What this world needs is a more equal distribution of wealth	U.S.	3.0	30.0	55.4	10.4	1.3
		France	20.3	24.9	24.8	28.8	1.3
Q123	We have gone too far in pushing equal rights in this country	U.S.	12.4	53.2	28.0	4.8	1.6
		France	24.7	30.3	18.2	24.9	1.9
Individualism							
Q116	I admire those who attempt to be independent and self-sufficient by growing their own food and adopting resource-conserving lifestyles	U.S.	.3%	6.5%	67.7%	24.9%	.6%
		France	7.0	16.9	29.7	45.9	.5
Q111	In a fair system people with more ability should earn more	U.S.	1.1	12.6	69.0	16.3	1.0
		France	8.6	15.6	30.6	44.7	.5
Q112	Government has no right to regulate people's personal risk-taking activities such as smoking, mountain climbing, hang gliding etc.	U.S.	2.6	23.9	53.4	19.2	1.0
		France	16.1	22.8	18.6	41.7	.9

distribution of wealth" (Q121). Moreover, two-thirds also disagreed that "We have gone too far in pushing equal rights in this country" (Q123).

A view known as individualism was measured in three questions. Strong support for this view was found in all three items. A vast majority of the American respondents (85.3%) agreed that "In a fair system people with more ability should earn more" (Q111). Nearly three-quarters (72.6%) indicated agreement with the statement "Government has no right to regulate people's personal risk-taking activities such as smoking, mountain climbing, hang gliding, etc." (Q112). Finally, there was almost complete agreement (92.6%) with the statement "I admire those who attempt to be independent and self-sufficient by growing their own food and adopting resource-conserving lifestyles" (Q116).

Two questions assessed the technological enthusiasm worldview (based on Jasper, 1990). A vast majority of the respondents (86.1%) thought that "A high-technology society is important for improving our health and social well-being" (Q108). However, 61.2% also agreed that "Technological development is destroying nature" (Q117).

An authoritarian/conservative worldview was assessed by two questions. Nearly three-quarters of the respondents were in favor of capital punishment (Q109). However, only 35.6% thought the police should have the right to listen to private phone calls to investigate a crime (Q125). Responses to a question designed to explore attitudes toward future generations were mixed. Respondents were divided over the statement in Q118, "Our technologies might impose risks upon future generations, but I believe future generations will be able to take care of themselves;" 49.6% agreed, 49.1% disagreed, and 1.4% were undecided.

Two questions examined attitudes toward economic growth. A large majority (80.2%) agreed that "Continued economic growth is necessary to improve our quality of life" (Q115). However, 43.8% also thought "continued economic growth can only lead to pollution and depletion of natural resources" (Q124).

FRANCE: WORLDVIEWS

French responses to the fatalism questions were mixed. Nearly 80% felt that they had very little control over risks to their health. However, a vast majority (87.5%) agreed that "People can offset health risks from pollution by improving their individual lifestyles, such as exercising and eating properly" (see Table 13). Only one-quarter indicated agreement to the statement "It's no use worrying about public affairs; I can't do anything about them anyway."

An overwhelming majority of the French respondents agreed with two of the hierarchical worldviews. There was almost complete agreement (89.5%) that people in positions of authority tend to abuse their power. Similar agreement was found in responses to the statement "Those in power often withhold information about things that are harmful to us." Over two-thirds of respondents agreed with the "need to pull together and support the energy choices our government has made."

French respondents tended to hold egalitarian worldviews. A majority of the French (81.6%) thought that "If people in this country were treated more equally we would have fewer problems." However, support was lower for the other two egalitarian measures. Only slightly more than half (53.6%) agreed that the world needs a more equal distribution of wealth. However, 55.0% disagreed that "We have gone too far in pushing equal rights in this country." General agreement with the individualistic view was found in all three questions. Three-quarters of the French respondents agreed that "In a fair system people with more ability should earn more." Similar agreement was found when respondents were asked whether they admired those with independent and resource-conserving lifestyles. Sixty percent felt that the government had no right to regulate people's personal risk-taking activities.

The French respondents expressed awareness of both the risks and benefits of technology. There was almost complete agreement (88.5%) that "A high-technology society is important for improving our health and social well-being." On the other hand, three-quarters of the respondents also agreed that "technological development is destroying nature." French responses to the conservative/authoritarian worldview questions were mixed. Respondents were divided over support for capital punishment; 53.9% were in favor of it and 45.3% were not in favor. Forty-seven percent agreed the police should have the right to listen to private phone calls to investigate a crime (52.6% disagreed).

Nearly three-quarters of the French respondents thought that although our technologies might impose risks on future generations, they believed future generations will be able to take care of themselves. A majority of the respondents agreed that "Continued economic growth is necessary to improve our quality of life." But the French also tended to believe that continued economic growth will lead to pollution and depletion of natural resources (68.4% disagreement).

U.S.–FRANCE COMPARISON: WORLDVIEWS

Table 14 displays the differences in responses between France and the U.S. Americans tended to express more agreement with individualistic worldviews. For example, U.S. respondents recorded 10.0% more agreement to the statement "In a fair system people with more ability should

Table 14: **WORLDVIEWS: U.S.–FRANCE COMPARISON**

	Difference
Q63 I feel that I have very little control over risks to my health	44.6%
Q115 Continued economic growth can only lead to pollution and depletion of natural resources	24.6
Q118 Our technologies might impose risks upon future generations, but I believe future generations will be able to take care of themselves	23.6
Q117 Technological development is destroying nature	13.5
Q125 The police should have the right to listen to private phone calls to investigate a crime	11.4
Q114 We need to pull together and support the energy choices our government has made	10.9
Q123 We have gone too far in pushing equal rights in this country	10.2
Q122 It's no use worrying about public affairs; I can't do anything about them anyway	7.5
Q110 People in positions of authority tend to abuse their power	4.2
Q126 Those in power often withhold information about things that are harmful to us	4.0
Q48 People can offset health risks from pollution by improving their individual lifestyles, such as exercising and eating properly	3.6
Q113 If people in this country were treated equally we would have fewer problems	3.0
Q108 A high-technology society is important for improving our health and social well-being	2.5
Q124 Continued economic growth is necessary to improve our quality of life	-.3
Q111 In a fair system people with more ability should earn more	-10.0
Q121 What this world needs is a more equal distribution of wealth	-12.2
Q112 Government has no right to regulate people's personal risk-taking activities such as smoking, mountain climbing, handgliding, etc.	-12.3
Q116 I admire those who attempt to be independent and self-sufficient by growing their own food and adopting resource-conserving lifestyles	-17.0
Q109 I am in favor of capital punishment	-19.2

Note. Values shown are differences in percent agree responses for France minus percent agree responses in the U.S.

earn more." U.S. respondents were more likely to think government has no right to regulate personal risk-taking activities, with 12.3% more agree responses than the French. Americans were also more likely to admire independent, resource-conserving lifestyles; they produced 17.0% more agree responses to this statement than the French. The item on which Americans had the greatest increase in agreement compared to the French was one endorsing capital punishment (a 19.2% difference).

French respondents were more likely to hold fatalistic views. In particular, French respondents recorded 44.6% more agree responses to the statement "I feel that I have very little control over risks to my health." The French were also more likely to believe the police had the right to listen to private phone conversations and that people should support their governments energy choices.

The French were less likely to express support for egalitarian worldviews. They were more likely to think their country had gone too far in pushing equal rights and were less likely to think that a more equal distribution of wealth was necessary. In addition, the French were more likely to feel that continued economic growth will lead to pollution and that technological development was destroying nature. They also were less likely to be concerned about future generations.

3.6 Personal and Demographic Characteristics

The final section of the survey asked questions about personal characteristics of the respondents, including social, economic, and behavioral information. This information is summarized below and related to ratings of risks.

U.S.: PERSONAL AND DEMOGRAPHIC CHARACTERISTICS

As mentioned above, the U.S. sample included 729 (48.2%) males and 783 (51.8%) females, with an average age of 39.7. Fifty-one percent of the respondents had two years of college or more; 40.8% had completed high school. Nearly 9.0% of the sample had not finished high school. The respondents were primarily white (84.3%), and most (81.9%) reported their personal health was good or excellent. Nearly half (47.6%) reported their annual household income as less than $35,000; 46.4% reported incomes over $35,000. In terms of political orientation, 29.7% considered themselves as liberal or very liberal; 33.4% thought they were middle of the road; and 35.3% indicated they were conservative or very conservative.

One about one-third of the U.S. respondents reported that they voluntarily participate in an activity that others consider a risk to their health or safety. When asked what activities they voluntarily participated in a range of activities were mentioned. Many of the activities were sports-related, such as skiing, scuba diving, hang gliding, bungee jumping, flying airplanes, and motorcycling. In addition, other types of voluntary risk-taking activities included drinking alcohol, smoking, and driving.

Four questions measured respondents' environmental activism during the past year. About three-quarters of the U.S. respondents reported that they had avoided using certain products that harm the environment (see Q147 in Table 15). Seventy percent indicated that they had purchased a higher priced product because it was better for their health or environmentally friendly (Q150). Only about half the respondents said they had voted or worked for candidates because of their positions on environmental issues (Q149). Less than one-quarter (22.2%) of the respondents indicated that they had been active in a group or that works to protect the environment (Q148).

Table 15: **ENVIRONMENTAL ACTIVISM**

Have you done any of the following things in the past year?		Have done	Have not done	Can't remember	Don't know/ no answer
Q147 Avoided using certain products that harm the environment	U.S.	74.4%	23.3%	2.0%	.3%
	France	64.7	33.9	.9	.5
Q148 Been active in a group or organization that works to protect the environment	U.S.	22.2	77.1	.5	.3
	France	9.1	90.9	0.0	0.0
Q149 Voted or worked for candidates because of their positions on environmental issues	U.S.	49.5	49.7	.3	.4
	France	26.5	72.8	.1	.6
Q150 Purchased a higher priced product because it was better for your health or environmentally friendly	U.S.	69.8	28.0	2.0	.3
	France	70.7	28.3	.6	.3

FRANCE: PERSONAL AND DEMOGRAPHIC CHARACTERISTICS

As mentioned above, the French sample included 729 (47.0%) men and 821 (53.0%) women, with an overall age of 42.2. Nearly two-thirds (63.6%) of the French respondents report monthly household incomes under 13,000 francs (which corresponds to about 31,200 U.S. dollars per year).

Two-thirds of the respondents reported their personal health to be good or excellent. In terms of political orientation, 30.6% of the respondents considered themselves left or far left, 18.1% middle of the road, and 24.9% right or far right. Only 21.5% of the French respondents reported voluntarily participating in activities that were risky to their health or safety. The most common type of voluntary activity included some type of sport; other types of activities included smoking, political activism, and going to demonstrations.

With respect to personal environmental activism, 70.7% of the French indicated that they had purchased a higher priced product because it was better for their health or environmentally friendly (see Table 15). Nearly two-thirds also reported that they had avoided certain environmentally harmful products. Only about one-quarter of the respondents had voted or worked for candidates because of their positions on environmental issues. Few French respondents (9.1%) reported being active in an environmental group or organization.

U.S.–FRANCE COMPARISON: PERSONAL AND DEMOGRAPHIC CHARACTERISTICS

American respondents were more likely to report voluntary participation in risky activities, with 35.4% of U.S. respondents reporting that they participate in such activities compared to 21.6% of French respondents. American respondents were also more likely to report voting for an environmental candidate in the past year, with 23.1% more "have done' responses than the French (see Table 16). In addition, Americans were more likely to report being active in an environmental group or organization.

Table 16: **ENVIRONMENTAL ACTIVISM: U.S.–FRANCE COMPARISON**

		Difference
Q150	Purchased a higher priced product because it was better for your health or environmentally friendly	.9 %
Q147	Avoided using certain products that harm the environment	-9.7
Q148	Been active in a group or organization that works to protect the environment	-13.1
Q149	Voted or worked for candidates because of their positions on environmental issues	-23.1

Note. Values shown are differences in percent agree responses for France minus percent agree responses in the U.S.

3.7 Summary

Many strong differences in attitudes and opinions of the French and American respondents were found. These differences center around issues of trust in government, science, and industry, and around certain worldviews such as egalitarianism and fatalism and not around perceptions of risk from nuclear power or nuclear waste. If anything, the French expressed more concern about risk from nuclear power and nuclear waste than did the Americans.

Across the entire set of hazards, the French were generally more likely than the Americans to call something a high risk. Nuclear power plants produced the smallest difference between the two countries in this regard. Americans were relatively more concerned about coal and oil burning power plants. Nuclear waste evoked much greater percentages of high-risk responses than did nuclear power and, in fact, France exceeded the U.S. in this regard.

There were also large differences between the French and Americans for the attitudinal statements. Trust stands out as a strong variable differentiating American and French respondents. Over two-thirds (67.9%) of the French respondents agreed that "decisions about health risks should be left to the experts," whereas only 23.2% of the Americans agreed with this statement, a difference of 44.7%. Higher trust levels in France were born out in other survey questions as well. Another large difference occurred in regard to the statement "I feel that I have very little control over risks to my health;" 79.3% of the French agreed compared to 34.7% agreement among the Americans.

The French were also much more likely than Americans to agree that:

- When there is a really serious health problem, the public health officials will take care of it.
- Our technologies might impose risks on future generations, but I believe future generations will be able to take care of themselves.
- We can trust the experts and engineers who build, operate, and regulate nuclear power plants.
- We need to pull together and support the energy choices our government has made.

Our results portray the French (relative to the Americans) as a people who believe that nuclear power has high risks but feel it is necessary for maintaining their economy and their standard of living. Although they believe they have little control over risks, they trust the experts and appear content to put their fate in the hands of the authorities.

In contrast, Americans are more likely to admire independence, respect individuality, dislike government intervention in personal risk-taking, and favor more equal distribution of wealth. Americans are also more likely to want public involvement in decision making. A surprising finding is that Americans were slightly more likely to agree that "the nuclear industry is capable of managing its wastes safely," although the degree of agreement was low in both countries (32.5% agreement in the U.S. vs. 26.6% in France).

Higher support was found in France for a new generation of nuclear power plants to supply the country's future electricity needs; 77.3% of the French respondents indicated support compared to 63.1% of the U.S. respondents. While there are high levels of support for nuclear reactors, the question is framed in such a way that support is contingent on the public's trust or confidence in the nuclear industry. The respondent is asked to assume the nuclear industry is correct that these new plants are safer.

In addition, as expected, many demographic effects were found in each country. Women in both countries tended to be more concerned than men not only about risks from nuclear power and chemicals but other types of risks as well (e.g., stress, AIDS, drinking alcoholic beverages, motor vehicle accidents). Women also tended to exhibit more concern about global and local environmental problems and were more likely to choose protecting the environment over ensuring an adequate supply of electricity. Men were more likely than women to support the building of new nuclear power plants than women and to trust experts and the government.

In general, older persons were more likely to rate a health risk as high, especially for non-nuclear types of risks (e.g., street drugs, storms and floods). Less-educated respondents were more likely to rate a health risk as high for all 25 hazards than college educated respondents.

4 CONCLUSIONS

The surveys described in this chapter have attempted to understand attitudes, perceptions, and behaviors that lie at the heart of public acceptance or rejection of nuclear power. The data from the two surveys clearly showed the greater concern and opposition to nuclear power among women. Similar results have been obtained in other studies as well (Flynn, Slovic, & Mertz, 1994). By examining the interaction between this gender effect and the racial composition of the respondents in the U.S. survey, Flynn, Slovic, and Mertz (1994 were able to gain insights into why women tend to be more anti-nuclear than men. They found no strong differences among nonwhite males, nonwhite females, and white females. Only white males stood apart from the others in having low perceptions of risk from all hazards and stronger support for nuclear power. Probing further, they found that 30% of the white males had extremely low perceptions of risk from nuclear power and other hazards. These low-risk-perception white males differed in having higher incomes and education (though these did not explain why they differed in their perceptions). More importantly, they also held highly authoritarian and anti-egalitarian attitudes. They believed more than others that decisions about nuclear power should be made by experts, and they were more likely than others to trust experts.

The U.S. and French surveys also demonstrated the strong influence of worldviews on perception and acceptance of nuclear power. Using this same survey data, Peters and Slovic (1996) found that persons endorsing individualistic attitudes or attitudes depicting a hierarchical social system tended to be pro-nuclear. Those favoring a more egalitarian society, in which power and wealth are more evenly distributed, tended to be anti-nuclear.

These worldview and race/gender results point toward a dramatic conclusion. Attitudes toward nuclear power are conditioned by the interplay of psychological, social, cultural, historical, and political factors that will not easily be changed by public information or educational campaigns. Our support or opposition to nuclear power is part of who we are and how we feel about society and our place in it. In the U.S., white males of conservative ideology tend to see nuclear power as safe and acceptable. Those of egalitarian persuasion and those who feel disenfranchised and alienated from society see nuclear power as dangerous and unacceptable. Our experiences, sometimes filtered through worldviews, contain an affective quality that also conditions our reactions to nuclear power.

A unique feature of the present study has been a comparison, using the same survey, between public views in the United States, where nuclear energy is resisted, and public views in France, where nuclear energy appears to be embraced. We found, to our surprise, that concerns about nuclear power and nuclear waste as measured by the survey were high in France and were at least as great there as in the U.S. Thus, even though perception of risk is a strong predictor of attitudes toward nuclear power within both the U.S. and France, it cannot account for the different level of reliance on nuclear energy in the two countries. Further analysis of the survey data uncovered a number of differences that might be central in explaining the difference between France and the U.S. Specifically, the French:

- saw greater need for nuclear power and greater economic benefit from it
- had greater trust in scientists, industry, and government officials who design, build, operate, and regulate nuclear power plants
- were more likely to believe that decision-making authority should reside with the experts and government authorities, rather than with the people.

These findings point to some important differences between the workings of democracy in the U.S. and France and the effects of different "democratic models" on acceptance of risks from technology. The French model involves less public participation and more centralized control. Political scientists have recognized that, in a climate of strong distrust, the French approach, in which policy formation and implementation is not accessible to public intervention, is expedient (Morone & Woodhouse, 1989). Campbell (1988) for example, argues that formal democratic institutions providing political access to nuclear critics may be fundamentally incompatible with commercial success of nuclear power.

What has historically worked in France, however, is unlikely to be achievable in the U.S. The French nuclear power program is run by the state, not private industry. Electricité de France has long had a strong reputation for being competent and putting service above profits. The French have a tradition of looking to a scientific elite for guidance in policy matters. Jasper (1990, noting that the word as well as the image of a "technocrat" arose in France, observed that "perhaps no other political system provides as large a role for people to exercise power on the basis of technical training and certification" (p. 83).

It should be noted, however, that things are changing in France. Regulators point out that there has long been a special demand upon nuclear technology, among all the energy producing technologies, to display signs of being highly controlled (Poumadère and Mays, 1997). The classical top-down model of decision still so salient to outside observers, is seen by those same regulators as becoming a "thing of the past". This decision model and its technocratic, even paternalistic profile have recently been challenged in two areas. One area is that of public information and access to decision rationales: the Chernobyl cloud, which "stopped at the border" for France's radioprotection administrator in 1986, was certainly lethal for the old style of nuclear expertise and communication (Poumadère, 1991; Mays and Egouy, 1998; Mays, Richard et al., 1998). The domestic handling by French authorities of the Chernobyl accident leaves a legacy of mistrust, especially among French residents under the age of 35 today (Mays, Marris et al., 1997). The second area in which strong challenge to old models is seen is in the area of high level radioactive waste management (Mays and Poumadère, 1996). France is noted internationally as having made a foray into a radically new model of legislative and "public" participation in management decisions, but analysis shows that old French patterns of centralization have not disappeared (Barthe and Mays, 1998).

America, since Thomas Jefferson, has had a different approach to democracy, and it is not surprising that attempts to restrict citizens' rights to intervene directly in national risk-management policies have been vigorously opposed. One example is the unsuccessful attempt in Congress to strip the state of Nevada of its rights to issue environmental and safety permits for nuclear waste studies at Yucca Mountain (Batt, 1992).

Given that the classical French approach is not likely to be acceptable in the U.S., restoration of trust there may require a degree of openness and involvement with the public that goes far beyond public relations and "two-way communication" to encompass levels of power sharing and public participation in decision making that have rarely been attempted (Flynn, Kasperson, Kunreuther, & Slovic, 1992; Kunreuther, Fitzgerald, & Aarts 1993). Even this, however, is no guarantee of success (Bord, 1988; Nelkin & Pollak, 1979). In many situations, we may have to recognize that relationships are so poisoned that trust and conflict resolution cannot realistically be achieved in the short run. The bitter conflict over the proposed nuclear waste repository in Nevada is a prime example of such a situation. To preserve the form of democracy Americans value so highly, they will need to develop new ways to manage risks and work constructively in situations where they cannot assume that trust is attainable (Kasperson, Golding, & Tuler, 1992; National Research Council, 1996).

ACKNOWLEDGMENTS:

This study was primarily supported by the Alfred P. Sloan Foundation with assistance from the National Science Foundation under grant SBR 9422754. We wish to thank Leisha Mullican and Kari Nelson for their assistance in the preparation of the manuscript.

5 REFERENCES

Barthe, Y. & Mays, C. (1998). *High profile and deep strategy: Communication and information practices in France's underground laboratory siting process*. Riskpercom national case study on communication and radioactive waste management. Note technique SEGR 98-18. Fontenay-aux-Roses: IPSN.

Batt, T. (1992, July 23). Nevada claims victory in Yucca deal. *Las Vegas Review-Journal,* pp. 1A-3A.

Bord, R. J. (1988). The low-level radioactive waste crisis: Is more citizen participation the answer? In M. A. Burns (Ed.), *Low-level radioactive waste regulation: Science, politics, and fear* (pp. 193-213). Chelsea, MI: Lewis.

Buss, D. M., Craik, K. H., & Dake, K. M. (1986). Contemporary worldviews and perception of the technological system. In V. T. Covello, J. Menkes & J. Mumpower (Eds.), *Risk evaluation and management* (pp. 93-130). New York: Plenum.

Campbell, J. L. (1988). *Collapse of an industry: Nuclear power and the contradictions of U.S. policy.* Ithaca, NY: Cornell University.

Dake, K. (1991). Orienting dispositions in the perception of risk: An analysis of contemporary worldviews and cultural biases. *Journal of Cross-Cultural Psychology, 22,* 61-82.

Drottz-Sjöberg, B. M. (1993). Risk perceptions related to varied frames of reference. In P. Hubert & M. Poumadere (Eds.), *Proceedings of the Third Conference of Society for Risk Analysis Europe.* Paris: European Section of the Society for Risk Analysis, 1991.

Flynn, J., Kasperson, R., Kunreuther, H., & Slovic, P. (1992, Summer). Time to rethink nuclear waste storage. *Issues in Science and Technology, 8*(4), 42-48.

Flynn, J., Slovic, P., & Mertz, C. K. (1994). Gender, race, and perception of environmental health risks. *Risk Analysis, 14*(6), 1101-1108.

Freud, S. (1924). *Collected papers.* London: Hogarth.

Galton, F. (1880). Psychometric experiments. *Brain, 2,* 149-162.

Jasper, J. M. (1990). *Nuclear politics: Energy and the state in the United States, Sweden, and France.* Princeton, NJ: Princeton University Press.

Jenkins-Smith, H. C. (1993). *Nuclear imagery and regional stigma: Testing hypotheses of image acquisition and valuation regarding Nevada* [Technical report]. Albuquerque, NM: University of New Mexico, Institute for Public Policy.

Kasperson, R., Golding, D., & Tuler, S. (1992, Winter). Social distrust as a factor in siting hazardous facilities and communicating risks (Individual and collective responses to risk). *Journal of Social Issues, 48*(4), 161-188.

Kunreuther, H., Fitzgerald, K., & Aarts, T. D. (1993). Siting noxious facilities: A test of the facility siting credo. *Risk Analysis, 13*, 301-318.

Mays, C. & Egouy, R. (1998). *Mission Communication: Press office and related activities at France's Institute for Nuclear Safety and Protection (IPSN). Riskpercom crossnational case study on institutional risk communication.* Note technique SEGR 98-19. Fontenay-aux-Roses: IPSN.

Mays, C. & Poumadère, M. (1996). Uncertain communication: Institutional discourse in nuclear waste repository siting. In V. Sublet, V. Covello and T. Tinker (eds.), *Scientific uncertainty and its influence on the public communication process.* Amsterdam: NATO Advanced Scientific Workshop Series, Kluwer Academic Press.

Mays, C., Marris, C., Bonnefous, S. & Brenot, J. (1997). *Perceptions of risk in France in 1996: Before, during and after Chernobyl's tenth anniversary.* Report to the European Commission DG 12. Note technique SEGR 97-93. Fontenay-aux-Roses: IPSN.

Mays, C., Richard, V., Bonnefous, S. & Marris, C. (1998). *Risk reporting in the French print media at Chernobyl's tenth anniversary.* Note technique SEGR 98-50. Fontenay-aux-Roses: IPSN.

Morone, J. F., & Woodhouse, E. J. (1989). *The demise of nuclear energy? Lessons for a democratic control of technology.* New Haven, CT: Yale University.

National Research Council. (1996). *Understanding risk: Informing decisions in a democratic society.* Washington, DC: National Academy Press.

Nelkin, D., & Pollak, M. (1979, August/September). Public participation in technological decisions: Reality or grand illusion? *Technology Review,* pp. 55-64.

Peters, E., & Slovic, P. (1996). The role of affect and worldviews as orienting dispositions in the perception and acceptance of nuclear power. *Journal of Applied Social Psychology*, 26 (16), 1427-1453.

Poumadère, M. (1991). "The credibility crisis", in B. Segerstahl (ed.) *Chernobyl: A Policy Response Study*, Berlin: Springer Verlag.

Poumadère, M. (1995). Cultural factors in public acceptance of nuclear power, in *Uranium and nuclear energy: 1995*, Proceedings of the 20th international symposium of the Uranium Institute, London, 20-21 Sept.

Poumadère, M. and Mays, C. (1997). *Energy Risk Regulation in France.* Working Paper Nr. 89 of the Center of Technology Assessment in Baden-Würrtemberg. Stuttgart: Akademie für Technikfolgenabschätzung.

Poumadère, M., C. Mays, P. Slovic, J. Flynn & S. Johnson (1994). What lies behind public acceptance? Comparison of US and French perceptions of the nuclear power option, *Proceedings of the International Atomic Energy Agency meeting on The Nuclear Power Option*, 5-8 Sept., Vienna.

Poumadère, M., Mays, C., Slovic, P., & Mertz, C.K. (1996). Diversity in meaning: risks compared in France and the USA. In O. Renn (ed.), *Risk Analysis and Management in a Global Economy; Vol. 2 Risk Perception and Communication in Europe.* Stuttgart: Center of Technology Assessment in Baden-Württemberg.

Slovic, P. (1987). Perception of risk. *Science, 236,* 280-285.

Slovic, P. (1990). Perception of risk from radiation. In W. K. Sinclair (Ed.), *Proceedings of the Twenty-fifth Annual Meeting of the National Council on Radiation Protection and Measurements. Vol. 11: Radiation protection today: The NCRP at sixty years* (pp. 73-97). Bethesda, MD: NCRP.

Slovic, P., Flynn, J., & Layman, M. (1991). Perceived risk, trust, and the politics of nuclear waste. *Science, 254,* 1603-1607.

Slovic, P., Flynn, J., Mertz, C. K., & Mullican, L. (1993). *Health risk perception in Canada* [Report No. 93-EHD-170]. Ottawa: Department of National Health and Welfare.

Slovic, P., Kraus, N. N., Lappe, H., Letzel, H., & Malmfors, T. (1989). Risk perception of prescription drugs: Report on a survey in Sweden. *Pharmaceutical Medicine, 4,* 43-65.

Slovic, P., Layman, M., & Flynn, J. (1991). Risk perception, trust, and nuclear waste: Lessons from Yucca Mountain. *Environment, 33,* 6-11, 28-30.

Smith, K. (1988). Perception of risks associated with nuclear power. *Energy Environment Monitor, 4* (1), 61-70.

Wundt, W. (1883). Über Psychologische Methoden. *Philosophische Studien, 1,* 1-38.

3

CROSS-CULTURAL STUDIES ON THE PERCEPTION AND EVALUATION OF HAZARDS

Bernd Rohrmann
University of Melbourne, Melbourne, Australia

TABLE OF CONTENTS

SUMMARY

In a series of socio-psychological field studies, perceptions and subjective evaluations of risky activities and environmental conditions were investigated in several countries. The aim of this cross-cultural project is to analyze the cognitive structure of judgments about the magnitude and acceptability of risks to which individuals are exposed; to explore disparities between different societal groups; and to compare risk judgments across countries in which risk issues in general as well as particular risk sources (e.g., industrial facilities or natural hazards) have different salience.

In a first series of studies, data were collected in Germany (N=217), New Zealand (N=224) and Australia (N=272). In each country, four groups of respondents were defined: people with a "technological", "monetarian", "ecological" or "feminist" orientation. Participants were asked for judgments on 24 hazards (based on a taxonomy) according to 12 risk aspects (derived from a structural risk perception model).

In a second phase, a modified data collection was conducted in China (N=270), and that study was fully repeated in Australia (N=203). Regarding hazards, 12 previously used items and 12 new items were included. The sampling in both countries focused on 3 groups of students (i.e., Geography, Psychology, Engineering) and a group of scientists.

Data comparisons for countries, for societal or professional groups and for types of risks yield a complex picture. Cross-cultural disparities are evident in two ways: groups affiliated with a particular professional, cultural and political orientations differ considerably in their judgment and evaluation of hazards; and considerable cross-national variation in risk perception exists as well. It is also obvious that some hazards are perceived as either more perilous or less severe than epidemiological risk data would suggest.

The results demonstrate the strong influence of socio-psychological factors and the cultural quality of risk evaluations. The findings are significant for a better understanding of people's subjective risk appraisal and also societal risk controversies. They can be utilized for designing comprehensive risk information, communication and education programs within and across cultural contexts.

This cross-cultural project will be continued and extended, with data collections in Germany and Singapore completed and currently undertaken in Canada and Japan.

1 PROBLEM

1.1 Risk research - the socio-psychological approach

The examination and evaluation of risks used to be dominated by approaches such as "quantitative risk analysis", developed in the natural sciences and economics, with a strong focus on hazard models, probability estimates and fatality rates as the ultimate criterion. However, large discrepancies between expert risk assessments and laypeople's concepts of hazard magnitude and acceptance of risks (as well as disagreements among different experts themselves) became more and more obvious. Under these conditions it seemed necessary to complement 'technical' risk research by social-scientific approaches in order to expand how "risk" is conceptualized and to understand the "psychology of risk".

The study of risk perception - i.e., people's judgments and appraisals of hazards they are or might be exposed to - is at the core of this research domain (which in fact is "quantitative" as well). It was strongly influenced by the seminal work of Slovic, Fischhoff and Lichtenstein (e.g., 1980, 1982, 1985).

Within social-science risk research, three main lines might be distinguished:

- *The psychometric approach*: Research using multi-dimensional scaling procedures in order to analyze the structure of beliefs and values people hold regarding risks; this approach - usually looking at a set of hazards with regard to a set of pertinent risk aspects - has been widely applied (reviews in Slovic 1992 or Rohrmann 1995).
- *Risk as a cultural construct*: Risk is seen as a "social and cultural construction" (Johnson & Covello 1987) - not an 'objective' entity to be measured independently of the context in which hazards occur; its meaning is bound o societal distinctions (overview in Rayner 1992). The "cultural approach" is mainly based on a 'macro-sociological' perspective; empirical research is fairly rare).
- *Socio-psychological views on risk management*: The question of "acceptable" and "accepted" risk (Fischhoff et al. 1982, Fischhoff 1994) is seen as the core issue. If risk sources are controversial (e.g., large-scale/new technologies) or underrated (e.g., personal health hazards or car accidents), risk communication as well as conflict resolution become crucial tasks (Renn 1992a, Fischhoff, Bostrom & Quadrel 1993), and risk perception research is an important input.

The involved disciplines (psychology, sociology, political science, philosophy) share a view that "risk" should be treated as a culturally defined, multi-dimensional and context-dependent concept (cf. Rohrmann 1998).

1.2 Cross-cultural risk perception studies

Within risk perception research, *cross-cultural* studies have evolved as the major interest, questioning the concept that risk perception can be treated in terms of 'communalities' or 'universals'. Macro-sociological, anthropological and philosophical literature (e.g., Dake 1992, Douglas & Wildavsky 1982, Heimer 1988, Johnson & Covello 1987, Rayner 1990, Schwarz & Thompson 1990, Schuez 1990, Wildavsky 1995; cf. also Cvetkovich & Earle 1991) has been most influential in establishing the "cultural approach" of risk research and providing a new theoretical background for empirical risk perception studies. According to this perspective, the evaluative process of risk perception is determined by the norms, value systems and cultural idiosyncrasies of societies or societal (sub-)groups. Consequently each society or social group is likely to have its own specific set of risks believed to be of concern.

It is important to note that "cross-cultural" differences can be studied from two perspectives (cf. chapter 1 of this book): In *cross-national* studies, data from different nations (e.g., France versus Germany) or types of countries (e.g. industrialized vs. developing ones) are compared. In *intra-national* 'cross-group' comparisons, differences in risk evaluation between societal groups (defined according to social/political views or membership in interest groups etc.) would be analyzed. While the latter study type is rare, quite a number of empirical investigations have looked at cross-national distinctions. The research reported in this text combines the two approaches.

Some selected studies - which are relevant in the context of the present project - are listed in *Table 1* (for a full documentation and review see Rohrmann 1995). While the majority of risk perception studies is restricted to student samples, most of the authors listed in Table 1 have looked at several specified groups.

The table also identifies sample sizes, number of hazards included and the number of risk aspects presented for evaluating risk sources). For all of these studies, the "psychometric para-

Table 1: **SELECTED RISK PERCEPTION STUDIES**

Author(s) / Year	N	Groups	Hazards	Aspects
Earliest psychometric risk perception study <USA>				
SLOVIC / FISCHHOFF / LICHTENSTEIN 1978/1980	76/175	1	30/90	13/22
Studies in Germany				
WINTERFELDT / JOHN / BORCHERDING 1981	57/68	1	14	4
BORCHERDING / ROHRMANN / EPPEL 1986	80	4	24	11
OPWIS / MAY 1985	66	1	16	20
BURGEMEISTER / WEBER 1993	203	1	9	13
ROHRMANN 1994	137	4	24	11
Studies in Australia and New Zealand				
ROHRMANN 1991, 1996	278	8	24	11
ROHRMANN 1994	339	8	24	11
Studies in China				
JIANGUANG 1994	238	1	20	5
ROHRMANN & CHEN 1999	473	4	24	12
Other studies investigating group differences				
GOSZCYNSKA / TYSZKA / SLOVIC 1991 (Poland/USA)	140	4	40	1/15
GOULD / GARDNER / etc 1988 (USA)	1021	2	6	12
KUYPER / VLEK 1984 (The Netherlands)	220	7	10	12
MARRIS / LANGFORD / ORIORDAN 1996 (England)	131/70	3	13	9
NYLAND 1993 (Brazil/Sweden)	144/119	5	100/33	1/1
PILISUK / PARKS / HAWKES 1987 (USA)	429	3	10	1
SJÖBERG / DROTTZ-SJÖBERG 1991 (Sweden)	236	10	16	3
SJÖBERG / KOLAROVA / RUCAI 1999 (Bulgaria/Romania)	240/192	6	100/28	6
SOKOLOWSKA / TYSZKA 1995 (Poland/Sweden)	601/1488	9	6/11	2
TIEMANN / TIEMANN 1985 (USA)	42/361	6	30/17	9

Notes: *N* = Number of respondents; *Groups* = Number of (substantive) subgroups within the study's sample; *Hazards* = Number of risk sources to be judged; *Aspects* = Number of judgements per risk.
Studies listed in *Italics* are those discussed in this chapter.

digm" is the main approach; some authors (e.g., Marris et al 1996) utilize qualitative approaches as well. Cross-national comparisons are usually restricted to two countries.

1.3 Aims of the present project

The project "Cross-cultural Comparison of Risk Perception (CCR)" was started in Germany (in collaboration with K. Borcherding) and then continued in Australia, with further data collections in New Zealand and China (in collaboration with H. Chen). The objectives of the investigation are:

- to design the study according to a conceptual framework for hazards, risk aspects and respondents;
- to analyze the structure of judgments about the magnitude and the acceptability of hazards to which individuals are exposed and the underlying psychological factors;
- to specify the relevance of hazard characteristics for the perception and evaluation of risks;
- to compare risk judgments across cultural contexts, looking at both the influence of professional and societal orientations within a country and disparities between different countries.

The overall goal is to achieve a better understanding of the cognitive process of evaluating risks as well as conflicts about risk issues, and to gain findings which are useful for improving the communication about risk problems between the various involved parties.

The purpose of the present paper is to report on two separate but related cross-national sets of studies on risk perception, aiming at integrating the findings. In the next section, the theoretical background and the methodological approach of these studies will be described, followed by a description of the data collection in the different countries. In section 3, selected results on mean ratings, disparities between hazards, group and country differences and correlational analyses (including structural models) will be presented. Finally, some considerations about the meaning of risk perception research and its potential for applied objectives (such as risk communication) will be outlined.

2 RESEARCH APPROACH

2.1 Principal project design

The general approach is to measure relevant risk evaluation criteria (i.e., risk aspects) with respect to heterogeneous set of hazards (i.e., risk sources) and to collect such data for distinctive societal groups (combining an intra-national and cross-national approach). Each of the facets of the 'problem space', i.e., aspects, sources and respondents, are treated according to a specific taxonomy. Data collection is based on a standardized psychometric instrument.

2.2 Conceptual framework for risk judgments

Studying risk perception means, to look at the cognitive structure of people's beliefs, feelings and appraisals regarding hazards. The substantive basis of this project is the theoretical framework developed in Borcherding, Rohrmann & Eppel (1986) and Rohrmann & Borcherding (1985). In a first step, relevant concepts for the subjective evaluation of risky activities and residential conditions were selected; in step 2, structural relations between the variables representing the respective cognitions have been hypothesized. This framework is shown in *fig. 1*.

The two key aspects of evaluating risks are perceived risk magnitude and acceptance of risk. It is assumed that these variables are influenced by specific characteristics of the studied haz-

Figure 1: **FRAMEWORK FOR VARIABLES AND THEIR STRUCTURAL RELATIONS**

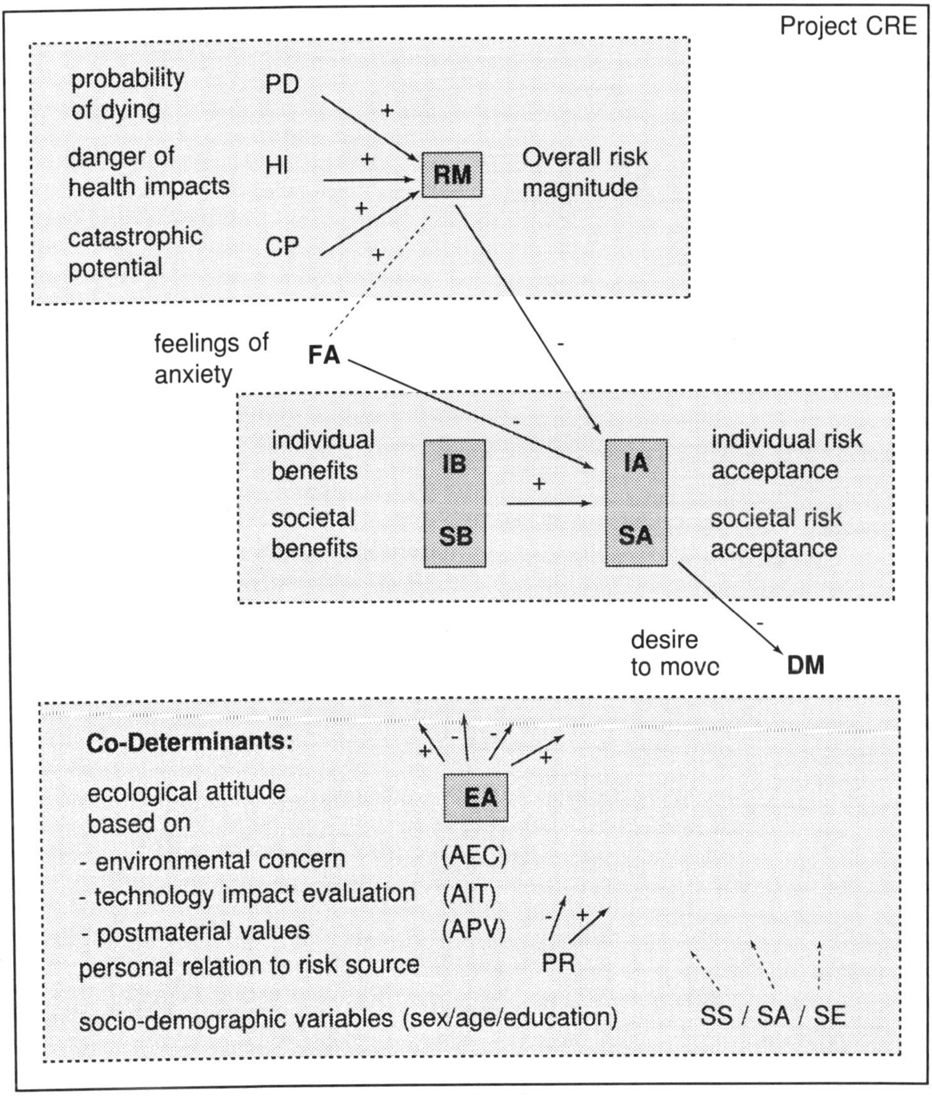

ards, negative impacts on the one hand and benefits on the other. Furthermore, ecological attitudes as well as socio-economic characteristics are introduced as relevant co-determinants. For both risk acceptance and benefits of risky enterprises, a distinction is made between the individual perspective and the societal viewpoint. This has to be considered in the operationalization of the concepts and the subsequent statistical analyses. The resulting list of variables is given in *Table 2.*

Table 2: **JUDGMENTAL ASPECTS FOR HAZARDS**

RM	=	Overall risk magnitude rating
PD	=	(Assumed) Probability of dying
HI	=	(Danger of) Health impacts
CP	=	Catastrophic potential (*)
FA	=	Feelings of anxiety about risk
AA	=	Attractiveness of activity (*)
IB	=	Individual benefit (of activity) (*)
SB	=	Societal benefit (of activity or hazard source)
IA	=	Individual risk acceptance
SA	=	Societal risk acceptance
PR	=	Personal relation/exposure to risk
DM	=	Desire to move (re risky residential conditions) (*)

(*) *these aspects do not apply to all hazards*

2.3 Hazards under study

As the risk type could have a strong influence on risk judgments, the selection of hazard to be investigated in this study was based on a taxonomy distinguishing four aspects: (1) activities/ professions vs residential/environmental conditions; (2) acute vs chronic impacts; (3) for activities: occupational vs private; (4) for residential conditions: natural vs technology-induced hazards. The resulting classification is shown in *Table 3.*

It should be noted that for each risk source the relation to humans was specified, e.g., "working as..." or "living near ..." while general terms (such as "cars", "asbestos", "nuclear power", "climate") were avoided.

In the first series of studies, each risk type was represented by three hazards (i.e., 24 altogether). This set-up was slightly changed (by exchanging some hazards) in order to adapt it for data collection in Asian countries.

2.4 Definition of samples: groups and countries

The cross-cultural agenda of this project is pursued in two ways: by comparing various groups within a society, and by including several countries.

GROUPS:

Sampling the population at large was not feasible, and taking just a convenience sample of local students seemed inadequate for the purpose of this project. The study has been based on a contrast group design instead. The rationale is to compare groups with specific societal, professional, and cultural orientations (in Dake 1991 or Douglas & Wildavsky 1982 the term

Table 3: **CLASSIFICATION OF RISK SOURCES**

	STUDY I	
	Activities	*Type/Impact*
A	Parachuting as a sport	private/acute
B	Driving in car races	
C	Down-hill ski training	
G	*Long-term heavy smoking*	private/chronic
H	Regularly taking tranquillisers	
I	*Eating too much and very fatty food*	
K	*Working as a fire fighter*	occupational/acute
L	Working as a blaster in quarries	
M	Flying an emergency service helicopter	
D	Working in asbestos production	occupational/chronic
E	*Working in an X-ray laboratory*	
F	Frequent working with air compressor tools	
	Residential conditions	*Type/Impact*
Q	Living in an avalanche-prone area	natural/acute
R	*Living in an earthquake-prone area*	
S	Living in an area w. many electric storms	
W	Living in a polluted dense urban area	natural/chronic
X	*Living in an area w. unhealthy climate*	
Y	Living in an area w. natural radioactivity	
T	Living near an explosives factory	technological/acute
U	*Living near a nuclear power plant*	
V	*Living near a chemical industry facilities*	
N	*Living near a coal power plant*	technological/chronic
O	Living near a smelter/metal prod. factory	
P	*Living near a large airport*	
	STUDY II	
	Physical hazards	
Z1	Cycling in dense urban traffic	private/acute
Z2	Regularly driving in cars	
C'	Swimming at dangerous beaches/rivers	
J'	Having unsafe/unprotected sex	private/chronic
I'	Intensive sun-bathing	
H'	Consuming hallucinogenic drugs	
L'	Working underground as a miner	occupational/acute
Q'	Living in an area prone to hurricanes	natural/acute
S'	Living in an area with frequent floods	
R'	Living in an area where often fires occur	
O'	Living near electric power lines/pylons	natural/chronic
X'	Living in an area with high air pollution	technological/chronic
	(*items in italics: employed in studies I **and** II*)	
	Sources of financial/social (non-physical) risk	
$2	Gambling some hours in a casino	financial/acute
$4	Regularly participating in sports betting	
$3	Being in places where thieves operate	
$1	Giving up a dissatisfactory but secure job	financial/chronic
$5	Investing in an uncertain product	
$6	Financing a new firm of unknown quality	

"worldview" is used). It is obvious from the political debates about risks, particularly those related to the use of technologies, that the professional socialization of the proponents leads to distinct evaluations of risk issues. (This notion relates to the "group" aspect of Douglas' "grid/group" concept of sociality; cf. also Thompson, Ellis & Wildavsky 1990). In the sampling plan, two factors were considered:

- Central values: in a first step, respondents with a predominantly '*technological*' or '*ecological*' orientation were studied; later respondents working in "*monetarian*" professions and those involved in "*feminist*" issues were included as further groups;
- Professional status: employed people and students were differentiated.

The societal groups included in the project were considered for the following reasons:
The literature on controversies about risks (especially those related to large-scale technologies) clearly shows that engineers/technicians on the one hand and environmental/'green' movements on the other hand are crucial proponents of this discourse (see, e.g., Conrad 1980, Gould, Gardner, DeLuca, Tiemann, Doob & Stolwijk 1988, Edwards & v. Winterfeldt 1987).

The groups sharing 'feminist' views (female employees being members of women's groups etc., and female students enrolled in a department of women's studies) were added because of findings that politically or socially involved women have specific views on technology, environmental issues and risks (see, e.g., Brody 1984, Fischer et al. 1991, Kistler & Jaufmann 1990, Savage 1993).

Finally, research on decision making and risk taking of managers (see, e.g., Keyes 1985, March & Shapira 1987, McCrimmon & Wehrung 1990, Shapira 1986) suggests that people regularly dealing with 'monetarian' issues have developed specific perspectives on risk evaluation and are less inclined to comply with 'ideological' positions.

In this grouping, gender is inevitably confounded with the ideological/professional factor; thus the possible impact of sex roles has to be considered.

COUNTRIES:

The countries included in the first sets of studies were Germany, Australia and New Zealand. These are all industrialized 'first-world' countries, but they differ in terms of their population and size, the significance of "nature" to the people, and the political significance of risk issues. The aim of the second set of studies is to compare risk perception in 'western' and 'eastern' nations. The present article contains a comparison of Australian and Chinese respondents (extensions by German, Canadian, Singapore and Japanese samples are underway).

It should be noted, however, that representative samples of these groups and countries were neither possible (due to limited resources) nor even intended. Certainly the project does not claim to compare "Germans" with "New Zealanders" or "Australians" with "Chinese". The idea, rather, is to use societal distinctions in order to elucidate typical intra-cultural and cross-cultural differences in risk evaluation.

2.5 Data collection - Study I

Study I consists of three sub-studies, one conducted in Germany <abbr. GER> (Borcherding, Rohrmann & Eppel 1986; Rohrmann & Borcherding 1985), a second one in New Zealand <NZL> (Rohrmann 1991/1996) and the largest one in Australia <AUS> (Rohrmann 1994). The initial German study was later extended to fully match the other two substudies. The samples achieved in the three countries (total N = 834) are listed in *Table 4*.

The places of data collection were: FRG: Mannheim, Darmstadt and München (all former West-Germany); NZ: Hamilton and Huntly; AUS: Melbourne and Brisbane.

Table 4: **SAMPLING: GROUP OF RESPONDENTS <STUDY I>**

Country: / *Sub-Group:*	AUS Australia	GER Germany	NZL New-Zealand
<T> „Technological orientation“	65	40	65
<T-e> Engineers	28	20	34
<T-s> Students of tech. sciences	37	20	31
<E> „Ecological orientation“	73	40	65
<E-e> Environmentalists	32	20	26
<E-s> Students of (env.) psychology	41+67	20	39+27
<F> „Feministic orientation“	72	60	67
<F-e> Members of feminist groups	40	30	47
<F-s> Students of women's studies	32	30	20
<M> „Monetarian orientation“	62	77	54
<M-e> Accountants/Finance managers	33	36	26
<M-s> Students in economics/finance	29	41	28
N = 834	272+67	217	251+27

Notes: „s“ denotes students, „e“ people employed in companies or institutions etc. Also studied: „Geography students“ in NZL, N=27; „Psychology-1 students“ in AUS, N=67; and „Experts in risk research“ (N=15; data not analyzed in detail because of small N).

In order to measure judgments about hazards (as outlined in the theoretical framework) as well as some personal characteristics of the respondents, a standardized instrument - the Hazard Evaluation Questionnaire HEQ - was constructed (originally in German).

The HEQ consists of three parts:

(1) judgments of 24 hazards according to 11 evaluative criteria;

(2) attitude scales; and

(3) questions about demographic variables.

Three attitudes were measured: environmental concern (AEC), evaluation of technology impacts (AIT) and societal value orientation (ASV) (economic vs. "post-material" values, Inglehart 1977). The items were taken from studies in environmental psychology (Fietkau, Hassebrauck & Watts 1980, Maloney, Ward & Braucht 1985, Prester, Rohrmann & Schellhammer 1987).

Note: In a few subsamples only, a fourth instrument was administered, dealing with attitudes toward risky decisions (ARD), i.e., risk propensity vs. risk aversion.

For all risk ratings, a response scale ranging from 0 to 10 was given, with 10 denoting the highest level of the criterion. The list of variables (altogether ca. 270 items) is identical for all groups of respondents.

In order to continue the project outside of Germany, an English version of the HEQ had to be prepared. No substantive changes were made apart from a few modifications to the instructional texts. The translation was thoroughly discussed with colleagues from Psychology and German Departments and checked via back-translations.

A brief note on the three countries:
With 7.7 mill. km^2, Australia is about 30 times as large as (former West-)Germany or NZ, but with 17 mill. inhabitants it is by far less densely populated than Germany or even NZ (however, 2/3 of the Australian population live in the 6 capital cities of the states). In political and societal terms, the three countries are principally similar in that they all adhere to democratic processes and 'western' values. Regarding industrialization and use of large-scale technologies, as well as in the importance of agriculture, Australia stands between NZ and Germany. Controversies about risk issues and "green" topics seem to be most salient in the German public debate. New Zealand appears to have less political/social tensions (at least until recently) than many other countries, and consensus is seen as a central social value. Within Australia, considerable cultural differences exist because of its much larger ethnic heterogeneity and disparities between the big cities and the rest of the nation. In Germany, the re-unification and integration of East- into West-Germany caused very substantial changes (note however that the data were collected beforehand). Generally, the three countries are exposed to the same kind of information about hazards and the risks they pose for humans.

In *Table 5*, selected demographic are given for all sub-samples. The mean attitude scores are in line with group definitions, e.g.: the group labeled as "Technological orientation" is least worried about (negative) impacts of technologies; environmental concern is highest for the "Eco-

Table 5: **SOME DEMOGRAPHIC CHARACTERISTICS OF THE RESPONDENTS**

Country:	**Australia**				**Gemany**				**New Zealand**			
Subgroup:	**Techn.**	**Mon.**	**Ecol.**	**Fem.**	**Techn.**	**Mon.**	**Ecol.**	**Fem.**	**Techn.**	**Mon.**	**Ecol.**	**Fem.**
Age	28.7	31.0	30.5	33.8	29.8	33.0	29.8	33.6	30.2	29.2	33.7	35.9
Sex (% Male/Female)	75/25	77/23	48/52	0/100	100/0	82/18	50/50	0/100	97/3	56/44	48/52	0/100
Employees/Students (%)	44/56	53/47	44/56	55/45	50/50	47/53	50/50	50/50	52/48	52/48	40/60	70/30
Attitudes (means) towards												
Impacts of technology (AIT)	2.9	2.9	4.3	4.0	2.7	2.8	4.6	4.3	3.1	3.2	3.9	3.9
Environmental concern (AEC)	4.3	4.1	5.3	5.2	4.3	4.8	5.8	6.1	4.6	4.4	5.3	5.2
Societal values (ASV)	4.4	4.2	5.4	5.4	4.6	4.2	5.6	5.5	4.7	4.2	5.4	5.3

Note: The attitude scores are based on 9/7/5 items which were measured on 7-point scales (1..7); where necessary, employees and students have been equally weighted.

logical" group; the non-material orientation within societal values is lowest for the "Monetarian" group; the attitudes of the "Feminist" group is very similar to the Ecological one.

2.6 Study II

The principal idea for Study II is to collect risk perception data in two 'western' and two 'eastern' countries. In phase <A>, two substudies were conducted, one in Australia <AUS> and one in China <CHI> (Rohrmann & Chen 1999). In phase <B>, data from Germany, Canada, Singapore and Japan will be added.

Due to research constraints, it was unfeasible to resemble the group setting of study I in China. Instead, mixed groups of "*scientists*" and "*students*" were targeted. Within students, *engineering/technology* and *psychology* students were sampled (as in study I); furthermore, a group of "*geography*" students was included. As there is more attention on natural hazards in study II (cf. Table 3 above), the risk perception of people trained in geography is particularly interesting. The data were collected in Beijing.

The Chinese group set-up was completely repeated in Australia (data collection in Melbourne). The available samples (total N = 473) are summarized in *Table 6*. (Note that the Australian samples from studies I and II are separate ones).

Table 6: **SAMPLING: GROUP OF RESPONDENTS <STUDY II, PHASE A>**

		AUSTRALIA	CHINA	total
Students				
T-s	Engineering/Technology	60	90	150
G-s	Geography	50	52	102
P-s	Psychology	60	74	134
Scientists				
X-e	Technical & Social Sciences	33	54	87
Sum		203	270	473

Note: In phase B, this study will be extended by sampling homologous groups (each N=60) in Japan, Canada, Singapore and in Germany (data collection under way).

For the Chinese substudy, the Hazard Evaluation Questionnaire - HEQ - was translated into Chinese language. The translation was thoroughly checked via back-translations by three scientists. The main variables and all instructions are identical to the English version, however, the set of hazards was adapted to the Chinese context, and some recently debated risks (such as electric power lines, gambling) were added; cf. Table 3.

A brief note on the two countries:

China is 9.6 and Australia 7.7 million square kilometers in area, but China's population density is 60 times higher. More importantly, there are huge disparities in cultural, political and economic terms: China is very different in its fundamental social and religious philosophies, it has not adopted a democratic political system as most 'western' countries, and it is less industrialized and far poorer than Australia. The level of public debate about risks, safety and environmental issues found in the west is not fostered. China appears to have a rather rigid social hierarchy, where people are expected to respect their superiors without much question. Governmental censorship procedures, the low proportion of non-Chinese residents and the lack of opportunities to visit other countries or learn about them through media is highly likely to restrict the dissemination of foreign ideas as well as international scientific findings about risks and to reduce the ideological and political diversification which occurs in Australia's multi-cultural society.

Selected demographic characteristics for study II are listed in *Table 7*. Mean age and the female/male proportion of the Australian and Chinese samples is pretty similar. Regarding attitudes, the largest differences in technology skepticism, environmental concern and societal (non-material) values are between the groups of Geography students (highest scores) and Technology students (lowest scores). Interestingly, environmental concern is considerably lower for all Chinese groups.

Table 7: **SOME DEMOGRAPHIC CHARACTERISTICS OF THE RESPONDENTS**

Country:	Australia				China			
Subgroup:	**Psych.**	**Geog.**	**Tech.**	**Scient.**	**Psych.**	**Geog.**	**Tech.**	**Scient.**
Age	19.1	20.6	21.5	40.6	21.1	21.1	21.0	39.5
Sex (% Male/Female)	15/85	32/68	53/47	82/18	31/69	25/75	54/46	59/41
Students/Scientists (%)	84			16	80			20
Attitudes (means) towards								
Impacts of technology (AIT)	3.4	3.7	3.0	3.2				
Environmental concern (AEC)	4.5	4.6	4.2	4.6	2.8	2.7	2.7	2.4
Societal values (ASV)	4.8	5.2	4.7	4.8				

Note: The attitude scores are based on 9/7/5 items which were measured on 7-point scales (1..7). AIT and ASV were not included in the Chinese study.

Within study II, the Chinese data deck is fully comparable to the Australian one; comparability to study I is given for selected hazards and subgroups of respondents.

2.7 Propositions

Because of the three-dimensional structure of the 'problem space', hypotheses can refer to risk aspects and their contingencies, to types of hazards and to differences between respondents in terms of subgroups or countries. In the following the main propositions are summarized.

DETERMINANTS OF RISK EVALUATIONS:

In general it is hypothesized for the "cognitive structure of beliefs, feelings and value orientations which underlay the subjective evaluation of risks

- that qualitative risk aspects, rather than (estimated) fatality rates determine risk magnitude judgments;
- that the structure of risk evaluations is dependent on the type of the risk source;
- that attitudes towards environment, technology and society significantly influence risk ratings and risk acceptance.

Regarding particular risk aspects, the proposed influence structure is depicted in the theoretical framework (cf. fig. 1).

SPECIFIC RISK SOURCES:

With respect to the taxonomy of hazards (cf. Table 3), the following was predicted:

- Perceived risk magnitude is higher for occupational than private risks and for technology-induced risk rather than natural hazards;
- probability-to-die ratings are higher for acute risks, while ratings of health impacts higher for chronic risks;
- risk acceptance is lowest for occupational and technology-induced hazards;
- most negative risk judgments are expected for nuclear energy.

Note that no overall difference in RM is expected between chronic and acute risks and risky activities versus residential risks, as the selection of the 2x12 hazards was intended to be balanced in this respect.

RISK PERCEPTION IN CULTURAL SUB-GROUPS:

Regarding differences between the four 'orientations' considered in this project, it is assumed that

- respondents involved in "ecological" or "feminist" issues give more negative risk evaluations and less acceptance of risks than the "technological" group, and that the judgments of the "monetarian" people would fall in between these extreme positions;
- people actually exposed to a risk source show lower RM and higher IA.
- students give more critical risk evaluations than employees.

Comments: It is assumed that both professional and philosophical factors ('worldviews') are particularly strong within the "ecological" and "feminist" orientation at one end (e.g., beliefs about fundamental environmental and societal crises) and with the "technological" orientation at the other end (e.g., optimism about the availability and feasibility of technical solutions); the "monetarian" orientation is likely to hold a 'middle' position, as this group should be less bound to such ideological positions. In terms of the study subjects, it was assumed that students of psychology, geography and environmental psychology show similar risk judgments but differ from students of natural sciences.

DIFFERENCES BETWEEN COUNTRIES:

In general, it is expected to find cross-national differences in the appraisal of particular hazards rather than in the cognitive structure of the investigated risk judgments. Also, the differences in risk perception across societal groups were assumed to be similar for the countries looked at.
Disparities expected for study I and study II include (details in Rohrmann 1994, Rohrmann & Chen 1999):

- Risk acceptance (IA/SA): AUS > NZ > FRG and CHI > AUS.
  Reasons: Compared to Germany, the societal 'risk debate' and 'safety thinking' seem to be less predominant in the other countries.

- More tolerance for specific risk sources if these are part of cultural background or political values. Examples:
  Extreme sports: NZ~AUS > FRG; AUS > CHI
  Occupational hazards: D < AUS~NZ; AUS > CHI
  Health & lifestyle risks: AUS~NZ > FRG; AUS > CHI
  Financial/social risks: D < AUS~NZ; AUS > CHI
  Natural hazards: AUS~NZ > D; CHI > AUS
  Large-scale technologies: AUS > FRG > NZ; CHI > AUS
  Note: Here effects are unlikely to be homologous, e.g., nuclear energy is particularly opposed in NZ and AUS, while airports are much more seen as essential than in Germany; in China the striving for modernization and economic progress might overrun risk worries.
- Group polarization: NZ~AUS < D; AUS > CHI
  Reasons: cultural disparity in the desire for consensus in societal issues, plus political differences in the extent of pressure from those in power.
- Influence of attitudes: AUS < NZ < D; AUS > CHI
  Reasons: Differences in the significance of ideology-related attitudes towards technology-use and environmental issues (and the degree to which political parties are bound to opposing worldviews).

Of course these (preliminary!) considerations are not hypotheses in a strict sense but rather conjectures. (They mainly originate from informal discussions with social scientists who have some idea about the countries and societal groups looked at in this project).

3 RESULTS

3.1 Data analysis: Overview

Each of the two studies yields a 4-dimensional data box, with hazards (24, i.e., 2x2x2 sets of 3), judgmental aspects (11), respondents (4x2 groups), and countries (3 or 2). Consequently, the statistical analysis of the collected data is quite laborious and yields a very large volume of results.

It is impossible to cover the findings for all countries in this text; see Borcherding & Rohrmann 1986, Rohrmann 1991, 1994, 1996 and Rohrmann & Chen (1999) regarding the German, New Zealand, Australian and Chinese data.

In the following, mainly two types of results are presented: mean ratings and differences of group means for various sets of respondents (available for all hazards and risk aspects) <cf. sections 3.2 - 3.5; and analyses of the cognitive structure of risk evaluations, based on correlations among risk sources <3.6 - 3.7>.

It should be noted that these correlations can be (and have been) determined in several ways:

- *across risk sources*: the respective correlation matrix could be computed for each respondent or for mean data sets (either means of the whole sample or for means of specific subgroups);
- *across respondents*: an analogous matrix could be computed for each risk source or mean ratings based on defined sets of risks.

Demographic information about the groups was given above in Tables 5 and 7. The mean age is quite similar across groups and countries. However, the male/female proportion varies considerably because it is confounded with professional orientation (hence identifying gender effects on risk perception is difficult).

3.2 Judgments about risks and benefits of hazards

Firstly, means for two country samples are considered (merging the respective sub-groups): *Table 8* presents German results, i.e., mean judgments in 11 risk aspects for 24 hazards; in *Table 9*, corresponding Chinese results are given. *Table 10* shows overall means for selected high-risk or low-risk hazards.

The following hazards get particularly negative evaluations in terms of perceived risk magnitude <RM>, health impacts <HI> and (non-)acceptability <IA/SA>: long-term heavy smoking, drugs, working in asbestos production, living near a nuclear power plant, air pollution, and chemical industry facilities. These risk sources also induce significant fear associations <FA>. Regarding catastrophic potential <CP>, the highest rating clearly is for nuclear power. For the 'classical' risk indicator, probability of dying <PD>, judgments are high for smoking and drugs but also some other risk sources, e.g., car racing. An indirect expression of risk perception, desire to move <DM; study I only> is strongest for people exposed to large-scale technologies such as airports or nuclear power facilities. Generally, the least negative ratings are given for hazards such as electric storm areas (study I) or cycling (study II).

Regarding benefits, a positive individual benefit <IB> is seen for sporting activities (e.g., skiing); and for public service professionals (e.g., fire fighters) both individual and societal

Table 8 : **HAZARD RATINGS: MEANS FOR 11 RISK ASPECTS - GERMAN SAMPLE (N=217)**

RM = Overall risk magnitude rating
PD = (Assumed) Probability of dying
HI = (Danger of) Health impacts
CP = Catastrophic potential
FA = Feelings of anxiety about risk
IB = Individual benefit (of activity)
SB = Societal benefit (of activity)
IA = Individual risk acceptance
SA = Societal risk acceptance
PR = Personal relation to risk
DM = Desire to move

Var.	**OR**	**PD**	**HI**	**CP**	**FA**	**IB**	**SB**	**IA**	**SA**	**PR**	**DM**	
Hazard												
A	5.5	3.0	3.5		5.7	7.8	1.2	7.2	5.3	2.7		Parachuting
B	7.5	5.3	5.1		6.0	7.8	1.4	6.3	3.3	2.4		Car racing
C	5.4	2.3	4.4		4.5	7.9	1.7	7.2	5.1	3.0		Skiing
D	8.2	3.3	7.3		7.4	3.7	3.6	3.3	3.2	2.7		Asbestos production
E	6.3	2.0	5.8		5.6	5.5	7.2	5.3	6.7	3.3		X-ray lab
F	6.7	1.8	6.4		5.4	3.4	5.4	4.1	4.9	2.3		Compressor tools
G	8.4	3.2	7.7		6.5	4.3	1.0	4.3	1.5	5.8		Smoking
H	8.2	3.5	7.6		7.4	4.3	1.1	4.0	1.8	4.5		Tranquilizers
I	7.6	3.2	7.0		5.8	4.6	1.3	4.5	2.4	4.5		Overeating
K	4.8	3.1	3.6		4.3	7.4	8.9	7.9	8.9	2.4		Fire fighter
L	4.8	3.1	3.7		4.9	5.6	6.0	6.8	6.8	1.9		Blaster
M	4.1	2.9	2.6		3.9	8.1	9.0	8.2	9.2	2.1		Emerg. helicopter
N	5.9	1.9	4.8	5.1	4.6		6.6	4.0	4.4	4.9	5.9	Coal power plant
O	5.9	1.8	4.7	4.9	4.5		7.2	4.0	4.4	3.8	6.0	Metal production
P	5.8	1.3	4.6	5.5	4.7		6.6	3.7	4.0	4.4	7.5	Airport
Q	5.8	2.7	2.5	5.5	4.8			4.8	4.6	1.8	5.4	Avalanche area
R	6.3	2.9	2.6	7.9	5.6			4.8	4.4	2.1	6.0	Earthquake area
S	3.1	1.2	1.5	3.2	2.4			7.1	7.1	2.3	2.8	Electr. storms area
T	5.3	2.1	2.7	6.1	5.0		4.0	4.0	4.3	2.4	5.3	Explosives factory
U	6.8	2.4	4.8	9.6	7.0		5.0	3.3	3.5	6.3	7.4	Nuclear power plant
V	6.7	2.2	4.9	7.8	6.2		6.4	3.4	3.9	5.6	7.2	Chemical industry
W	6.8	2.4	5.6	7.0	5.5			3.6	3.7	7.3	7.3	Polluted urban area
X	5.4	2.4	4.3	5.0	3.8			5.1	5.2	6.0	5.3	Unhealthy climate
Y	5.3	1.7	4.0	4.8	3.9			5.2	5.2	3.7	4.6	Natural radiation
	6.1	2.6	4.7	6.0	5.2	5.9	4.6	5.1	4.7	3.7	5.9	*(Mean)*

Empty cells: Hazards not rated on these risk aspects.

Table 9 : **HAZARD RATINGS: MEANS FOR 11 RISK ASPECTS - CHINESE SAMPLE (N=271)**

RM = Overall risk magnitude rating
PD = (Assumed) Probability of dying
HI = (Danger of) Health impacts
CP = Catastrophic potential
FA = Feelings of anxiety about risk
IB = Individual benefit (of activity)
SB = Societal benefit (of activity)
AA = Attractiveness of activity
IA = Individ. risk acceptance
SA = Societal risk acceptance
PR = Personal relation to risk source

Var.	**RM**	**PD**	**HI**	**CP**	**FA**	**IB**	**SB**	**AA**	**IA**	**SA**	**PR**	
Hazard												
Z1	3.5	2.9	2.5		3.3	3.5	4.0	*3.8*	7.0	6.3	6.1	Urban cycling
Z2	4.5	4.1	3.3		3.9	4.5	4.8	*6.1*	7.6	6.9	1.8	Car driving
C'	7.2	6.4	4.6		6.6	7.2	1.9	*3.5*	3.3	4.0	1.7	Dang. beaches
G	6.6	5.7	7.7		6.3	6.6	2.1	*1.7*	2.6	3.2	3.1	Smoking
J'	7.6	5.5	7.2		7.0	7.6	2.1	*3.9*	2.3	3.0	1.3	Unsafe sex
I'	4.9	4.1	6.3		5.4	4.9	1.9	*3.5*	3.0	4.3	1.5	Sun-bathing
I	4.9	4.0	6.1		4.9	4.9	2.5	*4.2*	3.9	4.9	2.5	Overeating
H'	8.7	7.4	8.4		8.1	8.7	1.3	*4.4*	1.4	1.6	0.9	Halluc. drugs
K	5.1	5.0	4.7		5.2	5.1	8.2	*5.3*	5.8	8.1	0.9	Firefighting
E	6.0	4.8	6.2		5.9	6.0	7.6	*3.7*	5.4	7.7	1.1	X-ray lab
L'	6.4	5.7	5.9		6.4	6.4	7.8	*2.8*	4.8	7.6	1.1	Undergr. miner
$3	2.8		2.6		3.6	2.8	4.8	*6.5*	6.7	6.4	2.0	Giving up job
$1	7.5		5.9		6.9	7.5	1.0	*4.3*	1.7	1.7	1.4	Gambling
$2	6.0		4.3		6.5	6.0		*2.2*	2.4	3.3	2.0	Thieve places
R	5.7	4.9	3.4	7.4	5.7				4.3	5.0	2.7	Earthquakes
Q'	6.2	5.0	4.0	6.3	6.0				4.0	5.0	1.5	Hurricanes
S'	6.2	4.9	4.2	6.2	5.7				3.7	5.0	2.1	Floods
X'	7.1	5.1	7.4	6.5	6.8				2.5	3.4	3.6	Air pollution
X	5.9	4.2	6.7	5.3	5.8				3.5	4.4	2.9	Unhealthy climate
P	6.2	3.8	6.4	4.9	5.7		7.6		3.4	4.6	1.0	Airport
N	5.8	4.0	6.1	5.3	5.5		7.3		3.8	4.9	1.2	Coal power plant
U	6.7	4.8	5.9	7.5	6.3		8.3		3.5	4.5	0.7	Nuclear power plant
R'	*6.7*	*5.6*	*4.9*	*6.5*	*6.5*				*2.9*	*4.1*	*1.4*	Fire areas
V	*6.2*	*2.8*	*4.7*	*6.1*	*5.7*		*7.4*		*5.2*	*4.3*	*1.3*	Chemical industry
	6.0	4.9	5.4	6.2	5.8	5.8	4.6	*4.1*	3.9	4.8	1.9	*(Mean)*

The 24 hazards appear in the same order as in the questionnaire. Empty cells are not measured. Scores in italics: These items were not included in the Chinese study; Australian results are given for illustration, instead.

Table 10: **EVALUATION OF HAZARDS: SELECTED OVERALL MEANS**

Judgments: Individual risk acceptance = IA; Societal benefit (of activity) = SB; Overall risk magnitude rating = RM

	RM	SB	IA
Occupational and private activities			
Long-term heavy smoking	8.3	1.4	4.4
Drugs (taking tranquillisers/using hallucinogenics)	8.1	1.6	4.0
Working in asbestos production (x)	8.4	3.3	3.3
Urban cycling (xx)	4.9	5.2	6.9
Residential/environmental conditions			
Living near a nuclear power plant	7.2	5.4	3.5
Living in polluted urban area / air pollution (x/xx)	6.8	-	3.9
Living near chemical industry facilities	6.6	6.4	4.0
Living in an area with many electric storms (x)	4.1	-	6.4

Notes: Means from all 5 samples, equally weighted (Study I = German, Australian and New Zealand data; Study II = Australian and Chinese data); overall N = 1241. Response scale: 0-10. „x" indicates hazard used in Study I only, „xx" = Study II only.

Figure 2: **RISK PROFILE OF HAZARDS**

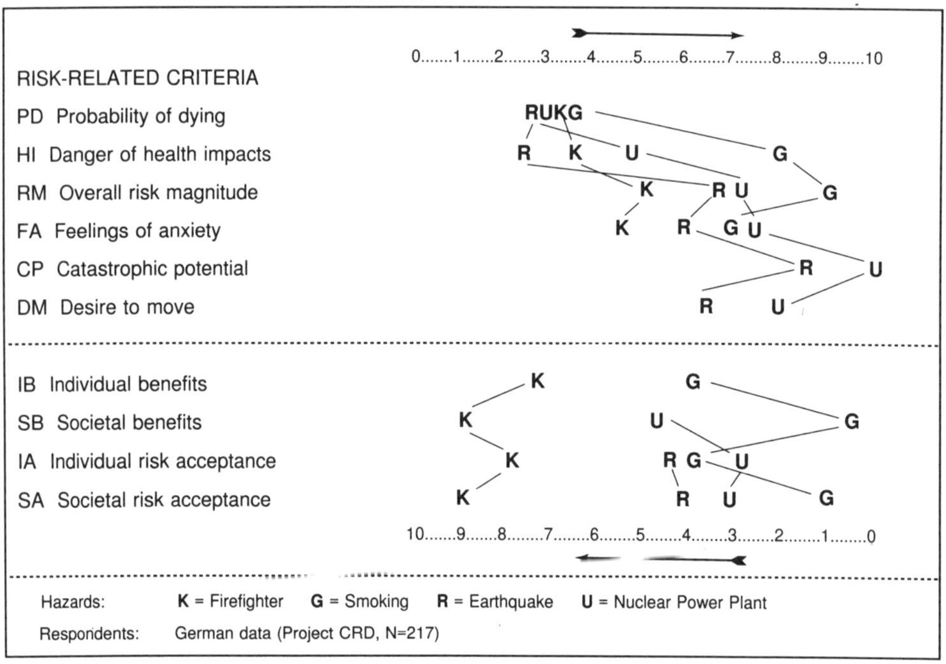

benefit <SB> are highly valued. Commonplace technical facilities (e.g., airports, metal production, coal power plants) are accepted as fairly beneficial as well. For a related aspect, the attractiveness of activities <AA; study II only>, scores are highest for car driving and lowest for smoking.

In *Figure 2*, the risk profile of four prototypic hazards is presented. The graph illustrates how differently such hazards are evaluated, once aspects other than probability of dying are considered.

Finally, the majority of respondents rate their personal risk level <PR> as low for most risk sources; main exceptions are hazards such as car driving, cycling, living in a polluted urban area, over-eating and smoking. Consequently hazard familiarity is unlikely to have a crucial impact on the judgments studied here (possibly except for driving and smoking).

3.3 Effects of hazard type

The set of hazards investigated in this project (cf. Table 3) was based on a taxonomy of risk sources, allowing for the comparison of defined hazard types. ANOVAs show significant effects of the four classification factors for most of the risk variables. The results - cf. *Table 11* for the German data - illustrate the following:

- Hazard impacts: Judgments of fatality rates are higher for risks comprising an acute danger (i.e., accidents/catastrophes); in comparison, health impacts are judged higher for chronic risk exposure.
- Benefits: for both occupational and private risky behaviors, people perceive benefits for themselves (even smoking or working with dangerous tools); however, benefits for the society relate to occupational activities only.
- Acceptance: Individual risk acceptance tends to be higher for private activities (e.g., sport or consumption risks), societal risk acceptance clearly is higher for occupational hazards. Re-

Table 11: **MEAN JUDGEMENTS FOR DIFFERENT TYPES OF RISK (German data, N=217)**

Type:	Kind of Hazard		Effect		Activities		Env. Cond.	
	Act.	Env.Cond.	Acute	Chronic	Private	Occup.	Natural	Technol.
n=	12	12	12	12	6	6	6	6
Risk Aspect:								
PD Probability of Dying			2.7 *	2.4				
HI Health Impacts			4.4	* 4.9				
IA Indiv. Risk Acceptance	5.8 *	4.4			6.4 *	5.4	5.1 *	3.7
SA Societal Risk Accept.	4.9 *	4.6			3.2	* 6.6	5.0 *	4.1
IB Individual Benefit					6.1 *	5.6		
SB Societal Benefit					1.3	* 6.7		

Notes: * indicates significant difference; „n" = number of hazards for that type; Hazards: „Act." = activities, „Env.Cond." = environmental conditions

Table 12: **MEAN RISK RATINGS: COMPARISON GER/AUS/NZL, 3 ASPECTS, N=217/272/224**

Var.:	**RM** Overall Risk Magnitude			**SB** Societal Benefit			**IA** Individual Risk Acceptance			
Data:	GER	AUS	NZL	GER	AUS	NZL	GER	AUS	NZL	
Hazard:										
A	5.5	6.5	6.0	1.2	2.2	2.2	7.2	7.5	7.7	Parachuting
B	7.5	6.9	6.6	1.4	2.7	2.4	6.3	7.4	7.6	Car racing
C	5.4	5.4	5.1	1.7	3.0	2.6	7.2	7.8	8.0	Skiing
D	8.2	8.6	8.3	3.6	3.0	3.3	3.3	3.1	3.5	Asbestos production
E	6.3	6.1	5.7	7.2	7.2	7.3	5.3	5.0	5.4	X-ray lab work
F	6.7	5.9	5.9	5.4	6.0	5.6	4.1	4.9	4.9	Compressor tools
G	8.4	9.0	8.9	1.0	0.9	0.7	4.3	5.2	5.2	Smoking
H	8.2	8.0	8.0	1.1	1.6	1.6	4.0	4.9	5.0	Tranquillizers
I	7.6	7.4	7.6	1.3	1.4	1.5	4.5	5.6	5.6	Overeating
K	4.8	6.1	6.3	8.9	8.7	8.7	7.9	6.6	6.7	Fire fighter
L	4.8	6.0	6.0	6.0	6.0	5.8	6.8	5.6	5.7	Blaster
M	4.1	5.2	5.6	9.0	8.7	8.7	8.2	6.9	7.0	Emerg. helicopter
N	5.9	5.6	4.7	6.6	6.3	6.0	4.0	4.3	4.5	Coal power plant
O	5.9	5.6	4.9	7.2	6.8	6.5	4.0	4.2	4.5	Metal production
P	5.8	4.3	4.7	6.6	7.1	7.0	3.7	4.9	5.0	Airport
Q	5.8	7.0	6.7				4.8	5.1	5.2	Avalanche area
R	6.3	6.8	6.1				4.8	5.1	5.7	Earthquake area
S	3.1	4.5	4.7				7.1	6.1	6.0	Electr. storms area
T	5.3	5.8	6.1	4.0	4.4	3.7	4.0	4.0	3.9	Explosives factory
U	6.8	7.6	7.6	5.0	4.3	3.8	3.3	3.2	3.1	Nuclear power plant
V	6.7	6.8	6.6	6.4	5.9	6.1	3.4	3.6	3.8	Chemical industry
W	6.8	6.6	7.3				3.6	4.5	3.8	Polluted urban area
X	5.4	5.8	6.6				5.1	5.1	4.6	Unhealthy climate
Y	5.3	6.7	6.8				5.2	4.6	4.3	Natural radiation
	6.1	6.4	6.4	4.6	4.8	4.6	5.1	5.2	5.3	(Mean)

Notes: Data based on overall sample (i.e., sub-groups merged). For results of significance tests for countries cf. Rohrmann 1994.

garding residential environmental risks, risk acceptance is higher for natural hazards from both a societal and an individual perspective. On average, risky activities are more accepted than risks related to residential hazards - voluntariness might be the crucial factor for this difference.

Differences in risk magnitude are small and were neither expected nor even intended, given the selection rationale for the risk sources considered in this study.

3.4 Comparisons across countries

Firstly, means of risk judgments obtained from the German (GER), Australian (AUS) and New Zealand (NZL) data deck is contrasted for three risk aspects; see *Table 12*.

The cross-national comparisons (including results not listed in Table 11 because of space restrictions) yield interesting differences between the Australian and German respondents. For example, the respondents from the AUS groups reveal more acceptance <IA, SA> for sport-related risks (e.g., car racing or skiing) and unhealthy private behaviors (e.g., smoking, overeating), and they give lower risk ratings <RM, PD, HI, CP> for 'conventional' technologies (e.g., airports, coal power plants). In contrast, risk-exposed occupations (even those of high social benefit <SB>, e.g., fire fighting), environmental pollution and large-scale technology such as nuclear energy get more negative evaluations than from the German respondents. However, given the views on nuclear power in Australia, the latter difference is not as large as expected. The results from AUS and NZL are much the same, apart from some disparities regarding earthquakes, a hazard more familiar to New Zealanders. While most findings are in line with the propositions outlined above, this is not true for occupational and natural environment hazards where data are inconsistent.

Interestingly, there is not much difference in risk sensitivity between the AUS, NZL and GER groups: the general range of the responses is similar in the three data sets, and the overall mean of all risk ratings is nearly the same (e.g., 5.2 vs. 5.3 vs. 5.1 for <IA>). Thus the assumed overall difference - lower perceived risk magnitudes and higher acceptance with Australian respondents - was not found.

The next *Table 13* presents mean judgments from study II for Australian (AUS) and Chinese (CHI) respondents, differentiating between the student and the scientists samples. Looking at cross-national differences first, major disparities in risk perception are apparent.

- Chinese risk magnitude judgments <RM> are significantly lower for cycling and lifestyle health hazards (e.g., smoking, overeating), except for drugs; and clearly higher than the Australian means for gambling.
- There is only one large difference in perceived societal benefits <SB>, namely for nuclear power plants where the Chinese view is far more positive.
- Very large differences emerge for individual risk acceptance <IA>. The hypothesis had been, higher <IA> in China (based on the - perhaps naive - proposition that Chinese people are more tolerant to hardships and better prepared to endure adversity). However, with the exception of cycling, Chinese <IA> means are lower for all hazards. The overall means (bottom line in Table 12) demonstrate that this is a principal difference and not hazard-dependent. It should be noted that the overall means for risk magnitude <RM> and societal benefit <SB> do not differ. (The same is true for the other risk aspects not included in Table 12, such as fear associations, individual benefits and societal risk acceptance). This points at a general cultural difference: consenting to significant risks within one's life - whether in the personal or the occupational sphere - seems to be unacceptable for most of the Chinese respondents. Also, people in China are not likely to foster individualistic views but rather think in terms of

Table 13: **MEAN RISK RATINGS: SELECTED COMPARISONS AUSTRALIA/CHINA, N=203/270**

Var.:		**RM** Risk Magnitude				**SB** Societal Benefit				**IA** Individ. Risk Acceptance			
	Country:	AUS	CHI	AUS	CHI	AUS	CHI	AUS	CHI	AUS	CHI	AUS	CHI
	Group:	*Stu*	*Stu*	*Sci*	*Sci*	*Stu*	*Stu*	*Sci*	*Sci*	*Stu*	*Stu*	*Sci*	*Sci*
Hazard:													
Z1	Urban cycling	6.2	3.4	6.6	4.0	6.2	3.5	7.2	5.7	6.8	7.1	6.3	6.4
Z2	Car driving	3.9	4.4	4.5	5.1	3.9	4.6	3.7	5.3	7.9	7.7	7.8	6.8
C'	Dang. beaches	7.1	7.0	6.7	7.8	2.2	1.9	1.5	1.6	5.6	3.3	6.0	2.6
G	Smoking	8.8	6.6	8.9	7.1	2.0	2.1	1.3	2.0	4.7	2.8	5.2	1.9
J'	Unsafe sex	8.5	7.7	7.5	7.7	1.9	2.2	2.0	1.8	4.9	2.4	5.9	1.4
I'	Sun-bathing	7.7	4.9	7.5	5.2	2.3	2.0	1.6	1.4	5.3	3.1	5.4	2.1
I	Overeating	6.9	4.9	6.8	4.9	2.7	2.5	2.4	2.3	5.6	3.8	5.9	3.0
H'	Halluc. drugs	8.2	8.6	7.4	8.9	2.3	1.2	1.8	1.1	4.9	1.6	5.9	0.6
K	Firefighting	6.6	5.2	5.8	4.8	8.4	8.2	8.3	8.7	7.1	6.0	6.6	5.3
E	X-ray lab	5.3	6.1	4.7	5.7	7.2	7.5	7.6	8.1	6.7	5.5	6.6	4.8
L'	Undergr.miner	7.1	6.6	6.2	5.9	6.9	7.7	7.3	8.2	6.1	4.9	6.0	4.2
$3	Giving up job	3.7	2.9	3.8	2.7	4.3	4.7	4.5	5.0	7.7	6.7	7.9	6.7
$1	Gambling	5.9	7.5	5.6	7.4	3.0	1.1	1.8	0.6	5.7	1.7	6.3	1.1
$2	Thieve places	5.9	5.9	5.2	6.3					5.0	2.5	5.0	1.5
R	Earthquakes	6.8	5.6	6.3	6.4					6.1	4.5	6.6	3.2
Q'	Hurricanes	6.7	6.2	6.2	6.2					6.1	4.1	6.7	3.4
S'	Floods	6.3	6.1	5.6	6.3					6.2	3.9	6.2	2.9
X'	Air pollution	6.2	7.1	6.2	7.6					5.0	2.6	5.0	1.7
X	Unh. climate	5.5	5.8	5.5	5.7					5.7	3.5	5.8	3.0
P	Airport	4.0	6.0	4.8	5.5	7.5	7.5	7.1	8.1	6.1	3.5	5.6	2.8
N	Coal p.plant	5.4	5.9	5.4	5.4	6.9	7.2	7.2	7.9	5.3	3.8	5.4	3.2
U	Nucl.p.plant	7.4	6.6	5.9	7.2	5.5	8.3	4.6	8.6	4.5	3.6	4.8	2.5
R'	Fire areas	6.3	6.6	6.5	7.0					6.1	3.4	5.8	2.0
O'	Power lines	4.7		4.4		7.9		7.8		6.0		5.9	
V	Chem.industry	6.2		6.3		7.4		7.3		5.2		5.0	
(Mean)		6.3	6.0	6.0	6.1	4.9	4.5	4.7	4.8	5.9	4.0	6.0	2.3

Note: „Student" results based on equally weighted means of the psychology, geography and technology student samples. (For results of significance tests for countries cf. Rohrmann & Chen 1999)

their family; if a hazard threatens the family's welfare or status it might be intolerable in any case.

- The difference in individual risk acceptance <IA> is particularly high for gambling and consuming hallucinogenic drugs - two activities which have quite a history in China but are now branded as illegal and seriously penalized. (There might be a 'social (un-)desirability' effect here as well.)
- Expected differences regarding occupational hazards, large-scale technologies and natural environment hazards did not manifest: both risk and benefit perceptions are rather similar in both country samples.

As further illustration, *fig. 3* shows a scattergram according to risk magnitude <RM> and societal benefit <SB>. The dotted lines depict the differences between mean judgments for the Chinese samples (shown in italics) and Australian samples. Urban cycling and gambling are examples of hazards which are rated quite differently on both dimensions.

A more thorough discussion of Chinese/Australian differences in risk perception is beyond the scope of this article (however, cf. Rohrmann & Chen 1999).

Figure 3: **RISK MAGNITUDE AND SOCIETAL BENEFIT FOR EIGHT HAZARDS**

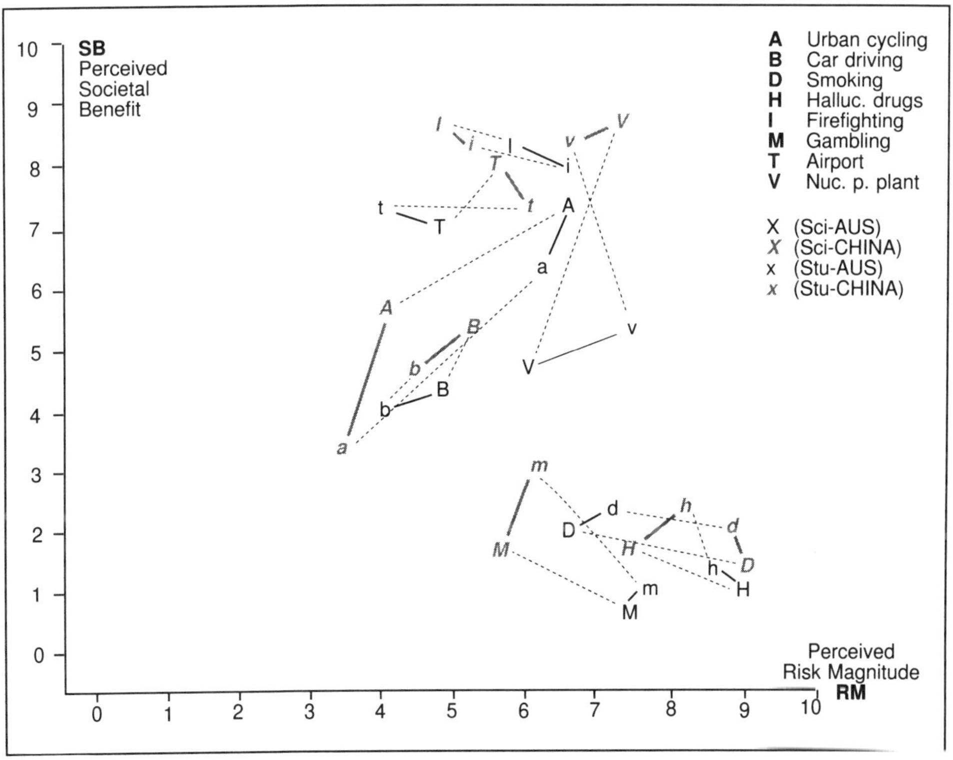

3.5 Disparities between societal groups

It was expected to find considerable differences between sub-groups. In *Table 14* selected means regarding three risk aspects are listed, using the German dataset. (Some results of significance tests for sets of risks, as well as "Eta" coefficients, indicating the magnitude of effects, are included in Table 15 below).

The findings of group comparisons can be summarized as follows:

- People with an "ecological orientation" as well as those involved in "feminist" issues evaluate risks much more critically than the other two groups. For example, for most risk sources, their ratings on all riskiness scales <RM, PD, HI, CP> and feelings of anxiety <FA> are

Table 14: **MEAN RISK RATINGS: GERMAN SUBGROUP DATA FOR SELECTED RISK ASPECTS**

Var.:	**RM** Risk Magnitude				**SB** Societal Benefits				**IA** Indiv. Risk Acceptance			
Group:	*Tec*	*Mon*	*Eco*	*Fem*	*Tec*	*Mon*	*Eco*	*Fem*	*Tec*	*Mon*	*Eco*	*Fem*
Hazard:												
A Parachuting	4.9	5.7	5.7	5.6	1.2	1.2	0.9	1.4	8.0	7.0	6.6	7.2
B Car racing	7.3	7.3	7.6	7.7	1.8	1.8	0.9	1.0	7.2	6.4	5.5	6.0
C Skiing	5.7	5.6	4.8	5.2	2.0	1.9	1.0	1.5	7.7	7.1	7.0	7.1
D Asbestos product.	7.5	7.9	8.0	9.0	4.6	3.3	3.9	3.2	4.1	3.8	2.6	2.7
E X-ray lab	5.1	6.0	6.5	7.4	7.4	7.3	7.0	7.2	6.0	5.7	4.7	4.5
F Compressor tools	5.7	6.5	6.6	7.8	5.6	5.5	5.6	5.1	5.2	4.7	3.1	3.1
G Smoking	8.0	8.5	7.9	8.7	0.8	0.9	0.9	1.3	3.4	4.0	4.0	4.5
H Tranquillisers	7.7	8.2	7.9	8.9	0.7	1.1	1.2	1.5	2.8	4.6	3.7	4.3
I Overeating	7.3	7.5	7.2	8.1	1.1	1.3	1.6	1.1	3.3	5.1	4.2	4.6
K Fire fighter	4.4	4.7	4.7	5.2	8.7	9.0	8.6	9.0	8.2	8.3	7.1	7.8
L Blaster	3.8	5.0	4.7	5.4	6.1	6.2	6.0	5.8	7.3	7.4	5.6	6.4
M Emerg. helicopter	3.5	4.3	3.9	4.6	8.8	9.1	8.8	9.3	8.3	8.4	7.8	8.2
N Coal power plant	5.0	5.7	6.2	6.8	7.1	6.3	7.4	6.1	5.5	4.3	3.5	2.8
O Metal production	5.1	5.6	6.1	6.7	7.8	7.1	7.3	6.9	5.1	4.4	3.5	2.9
P Airport	4.6	5.2	6.4	6.9	6.5	7.5	5.0	6.3	5.2	4.5	2.5	2.5
Q Avalanche area	6.0	5.7	5.1	6.2					5.0	4.9	5.3	4.1
R Earthquake area	6.2	6.3	5.8	6.8					5.3	5.0	4.8	4.1
S Elec. storms area	2.4	3.1	2.7	3.9					7.7	7.4	7.2	5.9
T Explosives factory	4.6	4.9	5.5	6.3	4.1	4.5	3.5	3.6	5.5	4.7	3.5	2.4
U Nucl. power plant	4.6	6.1	7.7	8.8	7.0	6.2	2.9	3.5	5.7	4.0	2.3	1.5
V Chemical industry	5.5	6.3	6.9	7.9	7.2	6.9	5.7	5.6	4.9	4.2	2.5	2.1
W Polluted urban ar.	5.4	6.7	7.2	7.6					4.6	4.0	2.9	2.7
X Unhealthy climate	5.1	5.0	5.7	6.0					5.5	5.7	4.6	4.5
Y Natural radiation	4.6	5.1	5.3	6.1					5.8	5.5	4.7	4.8
(Mean)	5.4	5.9	6.1	6.8	4.9	4.8	4.3	4.4	5.7	5.5	4.6	4.4

Notes: Data from students and employees merged within each of the subgroups.
For results of significance tests for subgroups cf. Rohrmann 1994.

higher, while benefit judgments <IB, SB> and risk acceptance <IA, SA> are lower. By comparison, those with a "technological orientation" show the lowest risk ratings, see more benefits and are more ready to accept risks. The judgments of the "monetarian" group fall in between the extremes. (*Note*: only RM, SB and IA are included in Table 14).

- This pattern is most obvious for technology-induced risks, as shown in *fig. 4*. The largest group differences occur with "living near a nuclear power plant" or "chemical industry" (which in fact are the most debated large-scale technologies anyway).

Figure 4: **JUDGMENTS OF TECHNOLOGICAL RISK BY FOUR GERMAN GROUPS**

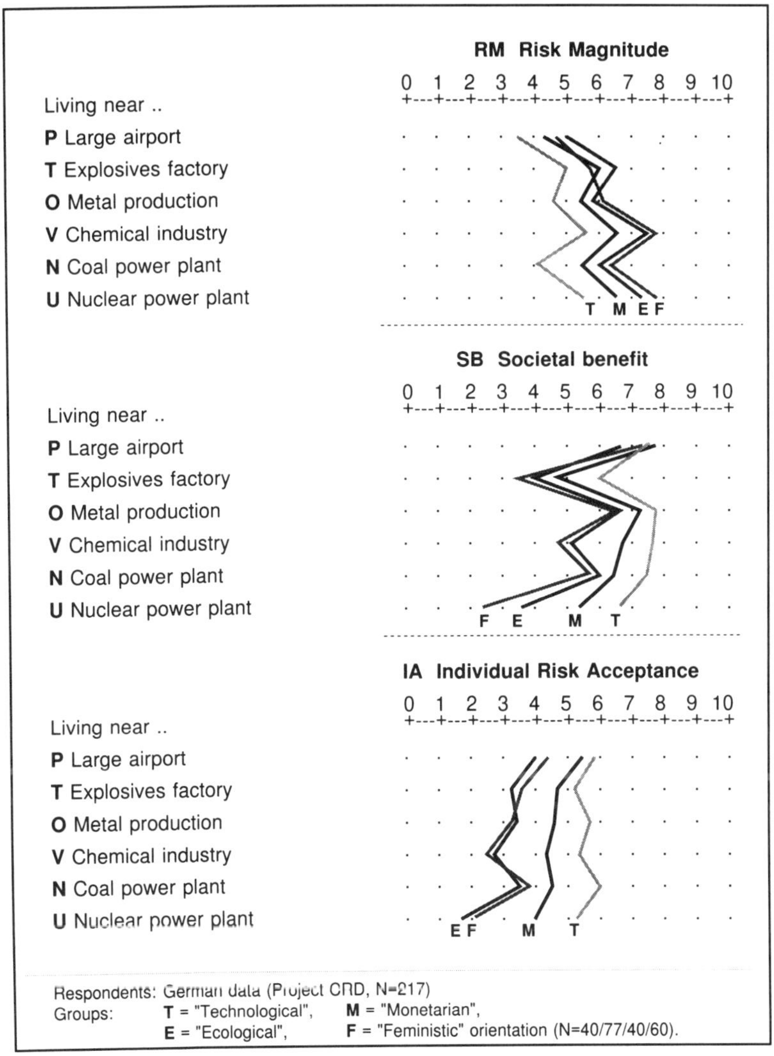

- There are some interesting exceptions from the general pattern. For example, with respect to 'consumption risks' such as smoking, tranquilizers and overeating, engineers and technology students give the lowest acceptance ratings while the "monetarian" and "feminists" groups yield surprisingly high scores.
- Group differences with respect to employees vs. students were also analyzed. The effects were smaller than expected and insignificant for most risk sources and risk aspects. Thus the groups were pooled.

How far are the described disparities among societal groups valid across countries? Respective comparative tables cannot be included here, but the findings compare as follows:

- The 'pattern' of disparities between groups with a "technological" or "monetarian" versus "ecological" or "feminist" orientation" is quite similar; most of the effects shown in Table 14 are true for the AUS and NZL data as well.
- Yet the described inter-group differences are stronger for the German than for the AUS or NZL data; the most salient examples are acceptability ratings and perceived benefits for technological risks such as X-rays, compressor tools, airports, chemical industry and nuclear power plants. This means that group polarization is lower for the Australian and NZL than the German groups, as it was predicted.

In *Table 15*, group and country effects were analyzed through two-way ANOVAs, using the datasets from three countries (AUS/NZL/GER) and two core groups (respondents with technological or ecological orientation). Eta-coefficients indicate the strength of effects. At least for the countries looked at here, disparities between societal groups tend to be stronger than cross-national differences, particularly regarding technology-induced risks.

Table 15: **SIGNIFICANCE AND ETA'S FOR MEAN DIFFERENCES BETWEEN GROUPS (Tec/Eco) AND COUNTRIES (AUS/NZL/GER) FOR TYPES OF RISK <ANOVAs>**

	RM		SB		IA	
Effect (Group, Country):	G	C	G	C	G	C
ABC Risky sport activities	-	-	***/29	**/19	-	-
DEF Dangerous work places	***/28	*/06	*/18	-	*/14	-
GHI Smoking/Pills/Overeating	-	-	-	-	*/17	**/14
KLM Dang. public-service jobs	**/19	***/37	-	-	-/	***/22
NOPTUV Technology hazards	***/43	-	***/41	-	***/32	-
QRS Natural hazards	-	***/28	(na)	(na)	-	-
WXY Unhealthy environments	***/24	***/21	(na)	(na)	*/10	-

Notes: Eta's given for significant effects only; decimal point omitted. All interactions CxG are insignificant, except of „IA" for sources NOPTUV (**). For abbreviations cf. Table 3, for number of cases cf. Table 5.

For the Chinese and Australian data, a subgroup of "scientists" is available; selected results are shown in *Table 16*, as well as subgroup means according to the subject area of students.

It appears that "geography" students are the most and "technology" students the least concerned about environmental hazards. Individual risk acceptance is lowest for the "scientists" group (particularly regarding lifestyle risks). In fig. 3 (presented above) differences between students and scientists are shown for both the Australian and the Chinese data. For most hazards, the student/scientist disparity is similar in both country samples (see, e.g., the judgments for cycling=A or smoking=D in the graph). Examples of reversed evaluations by students and scientists are two technological risks, airport and nuclear energy (see variables T and V in fig. 3).

Altogether the societal groups looked at in this research differ considerably and systematically in their risk perception.

Table 16: **MEAN RISK RATINGS: SELECTED CHINESE SUBGROUP DATA**

	Var.:	**RM** Risk Magnitude				**SB** Societal Benefit				**IA** Indiv. Risk Acceptance			
	Group:	*Psy*	*Geo*	*Tec*	*Sci*	*Psy*	*Geo*	*Tec*	*Sci*	*Psy*	*Geo*	*Tec*	*Sci*
Hazard:													
Z1	Urban cycling	3.8	3.6	2.9	4.0	3.0	3.9	3.7	5.7	7.5	6.6	7.2	6.4
Z2	Car driving	4.5	4.5	4.2	5.1	4.2	4.8	4.9	5.3	7.9	7.1	8.0	6.8
G	Smoking	6.4	6.9	6.4	7.1	1.9	2.0	2.3	2.0	3.3	2.2	2.8	1.9
J'	Unsafe sex	7.5	8.3	7.3	7.7	2.6	1.7	2.2	1.8	3.1	1.3	2.7	1.4
H'	Halluc. drugs	8.2	9.1	8.6	8.9	1.3	0.6	1.7	1.1	1.8	1.0	1.9	0.6
$3	Giving up job	2.6	3.0	3.0	2.7	4.2	4.6	5.4	5.0	7.0	6.7	6.4	6.7
$1	Gambling	7.1	8.1	7.4	7.4	1.0	0.6	1.6	0.6	2.2	0.9	2.0	1.1
$2	Thieve places	5.1	6.4	6.2	6.3					3.1	1.9	2.6	1.5
R	Earthquakes	5.5	6.3	5.0	6.4					5.1	3.6	4.8	3.2
S'	Floods	5.9	6.6	5.9	6.3					4.6	3.1	3.9	2.9
N	Coal p.plant	6.2	6.1	5.5	5.4	7.3	7.3	7.0	7.9	4.0	3.3	4.2	3.2
U	Nucl.p.plant	6.4	7.0	6.3	7.2	8.1	8.4	8.3	8.6	3.7	3.0	4.2	2.5
(Mean, 24 haz's)		5.9	6.3	5.8	6.1	4.5	4.4	4.6	4.8	4.4	3.5	4.1	3.2

Abbreviations: cf. Table 6. Empty cells: not measured.

3.6 Contingencies between judgments of hazards

The central research issue of this investigation is the cognitive structure of risk perception and evaluation. The propositions underlying the project are shown in the theoretical framework (see fig. 1 above). The assumed relationships were analyzed by means of multiple correlation and linear structural modeling.

As mentioned earlier, the correlations between risk aspects can be computed in several ways (cf. section 3.3 above). The matrices shown in *Table 17* are based on coefficients computed across hazards.

The results are mostly in line with the propositions stated earlier. As expected, correlations based on mean judgments - which do not reflect inter-individual differences - are principally higher. (Note that Table 17 is for the data set collected in Germany; the results for the data sets of the other countries are generally quite similar).

Table 18 presents selected single and multiple correlations for risk magnitude <RM> and individual risk acceptance <IA> as criteria, i.e., those computed across respondents.

Methodological notes: In this case, person-related variables, particularly attitude measures, can be included in the analysis, which is not possible if correlations across risk sources are used.

Table 17: **INTER-CORRELATION OF RISK ASPECTS VIA HAZARDS (GERMAN DATA)**

Above diagonal: correlation based on mean ratings;
below diagonal: Mean individual intercorrelation.

	RM	PD	HI	CP	FA	IB	SB	IA	SA	PE	DM	
RM	==	46	84	72	86	-58	-64	-69	-87	28	88	RM
PD	40	==	22	46	48	35	-48	18	-22	-23	14	PD
HI	64	34	==	30	67	-83	-45	-61	-67	33	80	HI
CP	46	31	31	==	95			-60	-74	36	64	CP
FA	60	42	51	62	==	-47	-60	-53	-70	13	75	FA
IB	-31	20	-46		-21	==	09	92	56	-36		IB
SB	-40	-27	-36		-32	13	==	18	71	-05		SB
IA	-47	-14	-43	-42	-39	54	27	==	81	-52	-93	IA
SA	-57	-18	-49	-46	-44	48	60	66	==	-42	-92	SA
PE	11	-02	16	20	05	-06	00	-10	-18	==	65	PE
DM	60	37	61	49	71			-68	-65	25	==	DM

Empty cells: Correlation cannot be computed. Coefficients are multiplied by 100. Variables CP, IB and DM are defined for 12 sources only; SB is defined for 18 sources (cf. Table 8).

The analyses were conducted for different sets of risk sources (which were aggregated as additive sum scores in case of sufficient homogeneity) and partly for subgroups as well. The results (for both the German data = Table 18 and the Australian and New Zealand data) demonstrate the following:

- As assumed, the (subjective) probability of dying <PD>, the concern about health effects <HI> and the catastrophic potential of risk sources <CP> (if applicable) are all predictive of perceived risk magnitudes <RM>.
- For hazards with the risk of acute impacts, such as sport activities (A/B/C) or natural hazards (Q/R/S), the probability of dying <PD> has higher weight than the health effects <HI>. However, for technological risks (sources N/O/P/T/U/V), health impacts <HI> is clearly a better predictor than the 'classical' risk indicator <PD>, and catastrophic potential <CP> is relevant as well.
- The proposed framework assumes that individual risk acceptance <IA> is significantly determined by risk magnitude <RM>, social benefit <SB> and feelings of anxiety <FA>. However, this model works well for technological risks only, with anxiety as the best predictor. Attitudes towards environmental issues, impacts of technology and social values (here aggregated into the overall ecological attitude <EA>) have considerable influence on both perceived magnitude and acceptance of technological risks.
- The main correlations are roughly similar for each of the considered subgroups (due to space limitations, results are given for one risk set only). Attitudes are less predictive though (possibly due to reduced variance in the case of intra-group analyses, particularly for the "Eco" subgroup).

There are two conclusions from these findings: firstly, that general attitudes are as influential as hazard-related considerations; and secondly, that the general pattern of psychological determinants of risk evaluations might not hold for specific hazards.

Table 18: **SINGLE AND MULTIPLE CORRELATION FOR RISK JUDGMENTS <German data>**

Criterion: RM = *Risk Magnitude*

		Predictors										
		PD Probability of Dying		HI Health Impacts		CP Catastrophic Potential		PR Personal Relevance		EA Ecological Attitude		
Haz. Set	*Group*	rPC	Beta	rPC	Beta	rPC	Beta	rPC	Beta	rPC	Beta	multR
ABC	T+E+M+F	.44	.45	.19	-.01			-.12	-.16			.46*
GHI	T+E+M+F	.36	.23	.35	.25			-.07	-.11			.42*
KLM	T+E+M+F	.37	.25	.36	.23			.06	-.03			.41*
NOPTUV	T+E+M+F	.52	.08	.62	.36	.53	.19			.46	.18	.67*
NOPTUV	Tec	.56	.26	.59	.37	.45	.18			.20	.16	.65*
NOPTUV	Eco	.57	.20	.65	.43	.48	.08			.37	.07	.68*
NOPTUV	Mon	.40	.02	.54	.40	.43	.15			.26	.15	.57*
NOPTUV	Fem	.46	.01	.62	.43	.50	.15			.46	.17	.65*
QRS	T+E+M+F	.40	.24	.37	.20	.24	.07			.17	.10	.45*

Criterion: IA = *Individual Risk Acceptance*

		Predictors										
		RM Risk Magnitude		IB Individual Benefit		SB Societal Benefit		FA Feelings of Anxiety		EA Ecological Attitude		
Haz. Set	*Group*	rPC	Beta	rPC	Beta	rPC	Beta	rPC	Beta	rPC	Beta	multR
ABC	T+E+M+F	-.16	-.15	.41	.43	-.05	-.13	-.01	.07			.45*
GHI	T+E+M+F	-.05	.00	.13	.11	.06	.03	-.09	-.07			.15
KLM	T+E+M+F	-.06	-.01	.14	.13	.04	.00	-.09	-.08			.16
NOPTUV	T+E+M+F	-.56	-.24			.49	.25	-.57	-.25	-.54	-.16	.70*
NOPTUV	Tec	-.30	-.23			.52	.42	.44	-.14	-.40	-.18	.65*
NOPTUV	Eco	-.51	-.47			.29	.29	-.45	-.24	-.18	-.30	.62*
NOPTUV	Mon	-.43	-.20			.41	.21	-.41	-.11	-.50	-.30	.61*
NOPTUV	Fem	-.53	-.23			.44	.30	-.56	-.35	-.47	-.17	.73*
QRS	T+E+M+F	-.18	.00					-.32	-.32	-.02	.03	.32*

Notes: rPC = single correlation between predictor and criterion; multR = multiple correlation; for abbreviations of risk sources cf. Table 3, for groups (and number of cases) cf. Table 4. * = multR significant ($p<.05$).

In addition to multiple regression, the LISREL approach (Jöreskog & Sörbom 1987, 1988) was used to analyze the structure of risk judgments. Its purpose is to identify "linear structural relationships" among constructs on the basis of a hypothesized theoretical model. The result of the analysis is shown in *Figure 5*. It was computed first using the German data (cf. Borcherding, Rohrmann & Eppel 1986; values on the left in the graph) and then the Australian data (Rohrmann 1994; values on the right). It is restricted to three risk sources, N/P/V = coal power plants, airports and chemical industries, which relate to three principal environmental problems, (air/noise and water pollution). The constructs included in the structural model represent the core of the framework in fig. 2; the measurement model is based on source-specific judgments (for <PD, HI, RM, SB, IA>) or on domain-specific attitudes (for <EA>). (Note that coefficients are multiplied by 100.)

Figure 5: **STRUCTURAL ANALYSIS OF RISK ASPECTS - MODEL <I>**

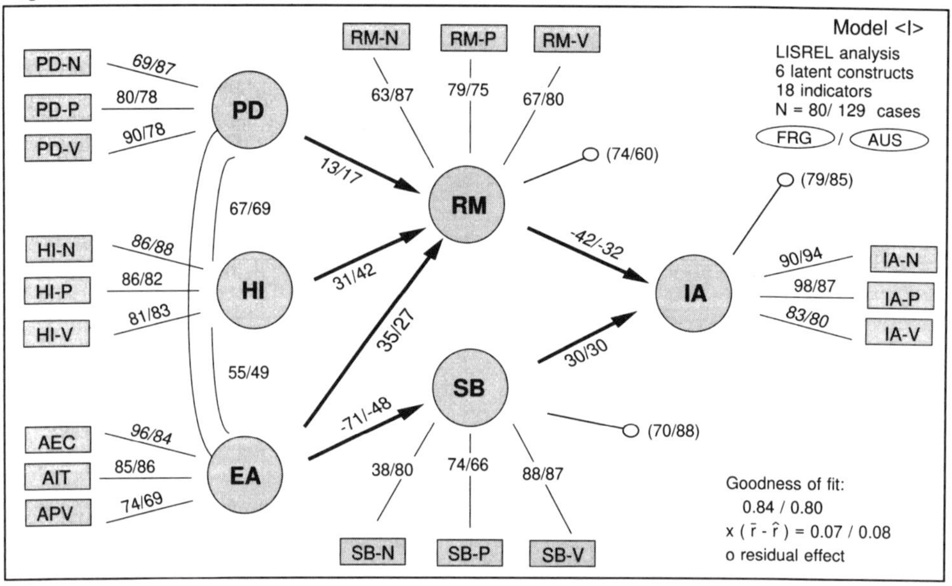

The results again demonstrate that, as predicted, risk acceptance <IA> is determined by perceived risk magnitude <RM> and benefit considerations <SB>; that concern about health impacts <HI> is more important than probability of dying <PD>; and that attitudes such as environmental concern, negative evaluations of technology and 'post-material' value orientations have considerable influence.

It should be noted that the "fit" of the model is only moderate (yet for comparability reasons the replicated model was not modified with respect to the path structure). The mean difference between the empirical correlation matrix and the respective model-derived coefficients is 0.08, indicating a fair correspondence.

Finally a higher-level LISREL model was developed by introducing overarching constructs, namely, "adverse impacts" (with probability of dying <PD> and concern about health impacts <HI> as indicators) and "risk as threat" (determined by risk magnitude <RM>, fear associations <FA> and catastrophic potential <CP>); these variables represent the negative evaluation of hazards. Societal risk acceptance <SA> is introduced as a final dependent variable. In terms of risk sources, as in some of the regression models above, six technological hazards (N/O/P/T/U/V) are used and directly aggregated into additive sum variables. The result (see *fig. 6*) is well in line with the basic claims of the proposed theoretical framework and it particularly affirms the significance of the "threat" aspect in risk perception.

Together these findings elucidate the psychological factors - such as negative associations or critical attitudes towards technology and its environmental and societal consequences - contributing to the 'intuitive' risk concept which people use for risk evaluations.

Figure 6: **STRUCTURAL ANALYSIS OF RISK ASPECTS - MODEL <II>**

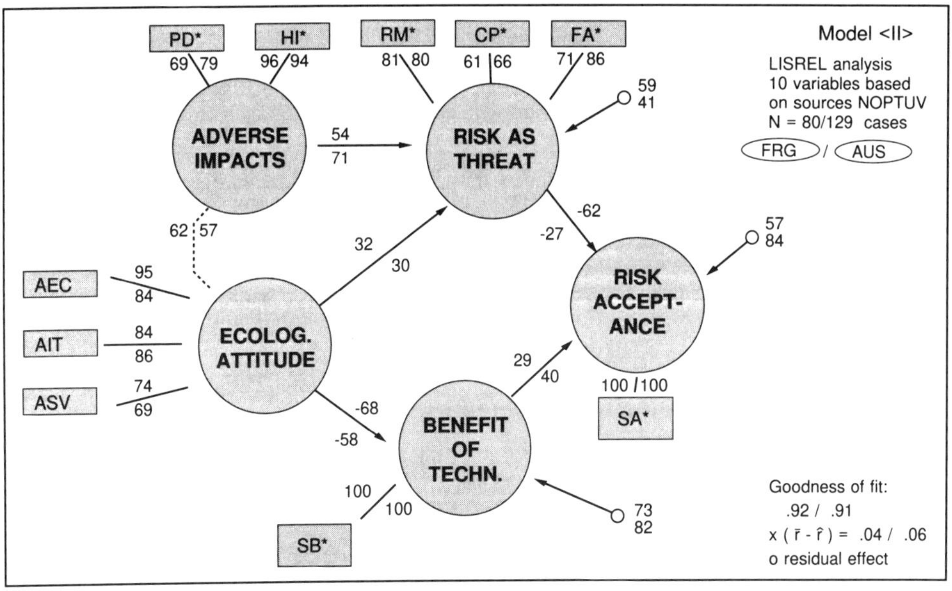

3.7 Cross-cultural differences in the structure of risk evaluations

Another issue is whether the identified contingencies between hazard judgments differ across groups and countries. Some societal subgroup comparisons have already been presented in Table 18. In terms of countries, the respective regression and LISREL models can be compared as long as the same group set is used. Results are available for study I (GER/AUS/NZL) but not yet for study II (AUS/CHI/GER).

As it can be seen from the LISREL model <I> (cf. fig. 5), the attained empirical structure is strikingly similar for the Australian and German respondents. The main disparity is that ecological attitudes <EA> are slightly less predictive for evaluations (both <RM> and IA>) of technological risks.

The results for model <II> (cf. fig. 6) again are quite homogeneous. There is one substantial difference: the concept of risk as a threat, developed within the Australian analyses and then applied to the previous data decks as well, is less predictive of societal risk acceptance <SA> in the German data. The main reason is that fear associations <FA> and catastrophic potential <CP> are less correlated with SA. The other parts of the model - and certainly the measurement structure - show similar coefficients.

In summary, similarities in the cognitive structure of risk evaluations - at least for the technology-related hazards looked at here - are more evident than disparities.

4 CONCLUSIONS

4.1 Interpreting cross-cultural risk perception studies

The two sets of cross-cultural studies reported here are evidence of the fact that *societal groups* affiliated with particular professional, cultural and political orientations differ considerably in their judgment and evaluation of hazards, and that considerable *cross-national* variation in risk perception exists as well. It is also obvious that some hazards are perceived as either more perilous (e.g., parachuting or nuclear power) or less severe (e.g., airports) than epidemiological risk data on health impairments and fatalities would suggest. Both findings are quite in line with the "cultural approach" to risk issues.

To understand the complex picture of risk perception, various psychological and social processes can be considered. In a theoretical framework developed by Kasperson & Renn (1992), the "social amplification of risks" is seen as a core phenomenon. They particularly stress the influence of factors such as political/environmental movements or media coverage, which either intensify or attenuate the perception and interpretation of risk-related events. The enormous (but quite selective) public attention to some risks over the last decade is indeed salient. Effects on individuals' cognition are not just distortions, however. Re-interpreting or modifying beliefs about risks serves various purposes, such as avoiding "cognitive dissonance" (Festinger 1964) when dealing with heterogeneous information or adjusting to one's peer group or deciding about the personal response to threatening but not-yet-present dangers. The more people are exposed to 'internal' conflicts or to 'external'/social influences (or even pressures), the more likely they will adjust their risk evaluations.

The notion of "worldviews" or "cultural biases" (i.e., culture-based orientations towards perceiving the world; Douglas & Wildavsky 1982, Thompson et al. 1990, Dake 1991) can certainly be applied to the societal groups looked at in the present study. (It might be tempting to relate this grouping to the prototypes postulated by the grid/group concept of cultural theorists, such as "individualism", "egalitarianism", "hierarchy", "fatalism". However, this would be completely speculative, as the conceptual background is too different; and the typology is not much supported by empirical data - cf. Marris et al. 1996, Sjöberg 1997).

Social psychologists have stressed that the individual history and the social context of people shape the personal set of values they apply to appraisals/evaluations of events and environments (e.g., Feather 1991, Maag 1991, Stern & Dietz 1994). Expressing and sustaining such values might seem to be harder in case of so-called alternative value orientations, e.g., 'ecological' attitudes. It has been found though that the views of minorities can have considerable influence on society if they maintain consistency and relate to evolving societal trends (cf. Moscovici 1985). The far-reaching influence of 'green' movements (at least in most 'western' countries) is a significant example of the impact determined societal groups can create.

However all the above considerations are based on social processes typical for 'Western' countries, so they might be applied to the Australian, New Zealand and German data of this project. It cannot be assumed though that the same is valid for countries which are less industrialized, less wealthy, less exposed to public debates and multiple independent media - such as many of the 'Eastern' nations.

Consequently, the comparative interpretation of findings from the Chinese data set is far more difficult because of the vast differences in political, economic and cultural terms (see, e.g., Hofstede 1980, Triandis 1989, and the handbook edited by Bond 1996). The eminence of social motivations and particularly family considerations seems to be the most important factor (stereotypically: Chinese collectivism versus Western individualism).

With regard to judgments and decisions, authors such as Yates & Lee (1996) conclude that risk is especially repugnant to Chinese decision-makers, that high-risk options are likely to be rejected and that risk aversion is common. This notion of cautiousness is in line with the notably low risk acceptance scores found in this project.

There are more factors to be considered though. Given the high power distance in Chinese society, the strong respect for status and authority, and also the apparent national pride, social-desirability effects might have influenced the risk judgments as well. In any case, far more research is necessary to achieve a sound understanding of risk perception in China.

Taken together, the outcomes of the presented cross-cultural studies clearly elucidate the crucial role of socio-psychological factors in the risk evaluation process. As has been discussed by various psychologists, sociologists and some engineers as well (e.g., Jungermann & Slovic 1993, Lopes 1992, Morgan 1993, Renn 1992b, Rohrmann 1995, Tweeddale 1994), the 'technical'/'quantitative' approach of risk analysis is inadequate to reflect the complex pattern of individual risk evaluations. The way people think about the magnitude and the acceptability of risks, and the way in which they make their respective judgments and decisions is influenced by knowledge, values and feelings. This includes a variety of 'qualitative' consequences of risky activities or living conditions that are not reflected in hazard scenarios, fatality probabilities and so on. Importantly, this general conclusion is valid across the different groups included in the investigation.

4.2 Considerations for further research

Cross-cultural risk perception research is a complex and diverse field of social-scientific study, and each single investigation is necessarily confined in many respects. In *substantive* terms, on-going research is needed for issues such as the following:

- For a comprehensive analysis of cultural influences on the interpretation of risks and risk acceptance, more social-psychological and sociological data are necessary (cf. McDaniels & Gregory 1991 for a research framework).
- In order to fully examine the relevance of intra-national versus cross-national cultural differences in risk perception, a wider range of both countries and societal groups needs to be reflected in the design of the studies (certainly student samples are not a sufficient basis for this issue).
- There is still a considerable lack of risk perception studies regarding Asian countries, and almost nothing is known with respect to South-American or African cultures (at least not in terms of internationally published research; cf. the documentation in Rohrmann 1995).

- The attitude-behavior link has not yet been systematically researched with respect to risk perception. How cognition and emotions about risks are related to actual behavior, e.g., risk-avoiding activities, job change, protests against risky technologies, and so on, needs to be thoroughly examined. Also, as it seems doubtful whether the same kind of contingency is valid for 'Western' and 'Eastern' cultures, cross-cultural studies would be particularly important.

From a *methodological* viewpoint, the empirical basis of the presented results is obviously small. Larger and more broadly defined samples are the most important means to increase validity. Another aim is to widen the scope of data collection approaches, in particular, to combine 'quantitative' and qualitative techniques - in order to get a richer understanding of how people feel and think about hazards and what their mental models (Bostrom et al. 1992) of hazards are.

Collaboration between researchers across countries would help to resolve these demanding tasks. The findings gained so far are certainly promising enough to justify continuing cross-cultural risk studies.

4.3 Relevance and utilization of findings

People are exposed to a multitude of hazards - as residents, at the workplace, as consumers of potentially harmful products or substances, when traveling, during leisure activities, and so on. All individuals have to deal with the related risks, and authorities have an array of responsibilities in terms of public health and welfare. Consequently, risk perception research is relevant to theoretical questions *and* practical issues.

This applies to risk information, risk education and risk communication, all cardinal elements of profound risk management (see, e.g., Covello, McCallum & Pavlova 1989, Fischhoff 1995, Kasperson & Stallen 1990, Lundgren 1994, National Research Council (USA) 1990, Rowan 1991, Viscusi & Magat 1987). Psychological expertise is indispensable for such tasks because effective information about hazards and related behavior modification programs require knowledge about existing cognitive and affective responses, and they are usually designed to change risk perception within a specific cultural context (Atman, Bostrom & Morgan 1994, Earle & Cvetkovich 1997, Rohrmann 1996).

Cross-cultural risk research in particular can help with problems such as risk information for ethnic subgroups (certainly in multicultural societies; cf. Vaughan 1995) and resolution of societal risk conflicts. Outcomes of social-science research can be utilized for a better understanding of controversies about risk evaluation and for improving risk communication among the various parties involved. Interactive communication and co-operative conflict management must be based on mutual knowledge and acceptance of the actors' way of thinking about the risks (Renn 1992b, Rohrmann 1991). In fact any kind of interaction and negotiation among people from different cultures - whether opponents or collaborators - would gain from awareness of cultural factors in risk appraisal and decision-making (Tse et al. 1988, Yates & Lee 1996).

To conclude, the approach adopted in this project - to investigate cross-cultural distinctions with respect to both societal groups and different countries - turned out to be constructive for investigating risk perception. Nevertheless, a wider range of cultures needs to be looked at in order to further clarify the influence of cultural factors on the cognition and evaluation of risks.

ACKNOWLEDGMENTS:
Parts of the German study were funded by grants from the German Research Society (Deutsche Forschungsgemeinschaft, DFG); for the Australian/Chinese study I received grants from Melbourne University and the Australian Research Council (ARC); and the Psychology Dept. of Waikato University (NZ) supported my project as well.

I also have to thank all the very many people who helped me conduct this research, including (to mention just three colleagues) Prof. K. Borcherding (Germany), Dr. P. Taylor (New Zealand), and Prof. H. Chen (China), and a little armada of research students - plus some 1300 respondents who were kind enough to participate in the project!

5 REFERENCES

Atman, C. J., Bostrom, A., Fischhoff, B., & Morgan, M. G. (1994). Designing risk communications: Completing and correcting mental models of hazardous processes (part 1). *Risk Analysis, 14*, 779-788.

Bond, M. H. (Ed.) (1996). *The handbook of Chinese psychology*. Hong Kong: Oxford University Press.

Borcherding, K., Rohrmann, B., & Eppel, T. (1986). A psychological study on the cognitive structure of risk evaluations. In B. Brehmer, H. Jungermann, P. Lourens, & G. Sevon (Eds.), *New directions in research on decision making* (pp. 245-262). Amsterdam: North-Holland.

Bostrom, A., Fischhoff, B., & Morgan, M. G. (1992). Characterizing mental models of hazardous processes: A methodology and an application to radon. *Journal of Social Issues, 48*, 85-100.

Brody, C. J. (1984). Differences by sex in support for nuclear power. *Social Forces, 63*, 209-228.

Burgemeister, J., & Weber, M. (1993). Risiko und Akzeptanz von Industrieansiedlungen. *Zeitschrift für Betriebswirtschaft, 63*, 147-169.

Chen, H., & Rohrmann, B. (1996). *Perceptions of risk of Chinese and Australian students and scientists*. Contribution to the International Congress of Psychology, Montreal.

Conrad, J. (Ed.) (1980). *Society, technology and risk assessment*. London: Academic Press.

Covello, V. T., McCallum, D. B., & Pavlova, M. (1989). *Effective risk communication. The role and responsibility of government and nongovernment organizations*. New York: Plenum.

Cvetkovich, G., & Earle, T. C. (1991). *Risk and culture*. Special Issue, Journal of Cross-cultural Psychology.

Dake, K. (1991). Orienting dispositions in the perception of risk - An analysis of contemporary worldviews and cultural biases. *Journal of Cross-cultural Psychology, 22*, 61-82.

Dake, K. (1992). Myths of nature: Culture and the social construction of risk. *Journal of Social issues*, *48*, 21-37.

Douglas, M., & Wildavsky, A. (1982). *Risk and culture: An essay on the selection of technical and environmental dangers*. Berkeley: University of California Press.

Earle, T. C., & Cvetkovich, G. (1997). Culture, cosmopolitanism, and risk management. *Risk Analysis*, 17, 55-77.

Edwards, W., & Winterfeldt, D. v. (1987). Public values in risk debates. *Risk Analysis*, *7*, 141-158.

Feather, N. T. (1991). Human values, global self-esteeem, and belief in a just world. *Journal of Personality, 59*, 83-106.

Festinger, L. (1964). *Conflict, decision and dissonance*. Stanford: Stanford University Press.

Fietkau, H. J., Hassebrauck, M., & Watts, N. (1980). *Der internationale Umweltfragebogen (IUF): Ein Instrumentarium zur Erfassung umweltbezogener Werte*. Berlin: Internationales Institut für Umwelt und Gesellschaft, Report IIVG/80.

Fischer, G. W., Morgan, M. G., Fischhoff, B., Nair, I., & Lave, L. B. (1991). What risks are people concerned about? *Risk Analysis*, *11*, 303-315.

Fischhoff, B. (1994). Acceptable risk: a conceptual proposal. *Risk: Health, Safety & Environment*, *5*, 1-28.

Fischhoff, B. (1995). Risk perception and communication unplugged: Twenty years of process. *Risk Analysis*, *15*, 137-146.

Fischhoff, B., Bostrom, A., & Quadrel, M. J. (1993). Risk perception and communication. *Annual Review of Public Health*, *14*, 183-203.

Fischhoff, B., Lichtenstein, S., Slovic, P., Derby, S. L., & Keeney, R. L. (1982). *Acceptable risk*. Cambridge: University Press.

Goszczynska, M., Tyszka, T., & Slovic, P. (1991). Risk perception in Poland: A comparison with three other countries. *Journal of Behavioral Decision Making*, *4*, 179-193.

Gould, L., Gardner, D., DeLuca, D., Tiemann, A., Doob, L., & Stolwijk, J. (1988). *Perceptions of technological risks and benefits*. New York: Sage.

Heimer, C. A. (1988). Social structure, psychology, and the estimation of risk. *Annual Review of Sociology*, *14*, 491-519.

Hofstede, G. H. (1980). *Culture's consequences*. Beverly Hills, CA: Sage.

Inglehart, R. (1977). *The silent revolution*. Princeton: University Press.

Jianguang, Z. (1994). Environmental hazards in the Chinese public's eyes. *Risk Analysis*, *14*, 163-169.

Jöreskog, K. G., & Sörbom, D. (1988). *LISREL (Analysis of LInear Structural RELationships) - User's guide*. Chicago: National Educational Resources.

Jöreskog, K. G., & Sörbom, D. (1987). *SIMPLIS - A simplified version of LISREL*. Mooresville: Scientific Software Inc.

Johnson, B. B., & Covello, V. T. (Eds.) (1987). *The social and cultural construction of risk*. Dordrecht: Reidel.

Jungermann, H., & Slovic, P. (1993). Characteristics of individual risk perception. In Bayerische-Rueck (Ed.), *Risk - a construct.* (pp. 85-102). München: Knesebeck.

Kasperson, R. E., Kasperson, J. X., & Renn, O. (1992). The social amplification of risk: Progress in developing an integrative framework. In S. Krimsky & D. Golding (Eds.), *Social theories of risk* (pp. 153-178). New York: Praeger.

Kasperson, R. E., & Stallen, P. M. (Eds.) (1990). *Communicating risks to the public*. Dordrecht: Kluwer.

Keyes, R. (1985). *Chancing it: Why we take risks*. Boston, MA: Little, Brown and Co.

Kistler, E., & Jaufmann, D. (Hg.) (1990). *Mensch - Gesellschaft - Technik*. Wiesbaden: Leske + Budrich.

Kuyper, H., & Vlek, C. (1984). Contrasting risk judgements among interest groups. *Acta Psychologica, 56*, 205-218.

Lopes, L. L. (1992). Risk perception and the perceived public. In D. W. Bromley & K. Segerson (Eds.), *The social response to environmental risk* (pp. 57-74). Dordrecht: Kluwer.

Lundgren, R. (1994). *Risk Communication*. Columbus: Batelle Press.

Maag, G. (1991). *Gesellschaftliche Werte. Strukturen, Stabilität und Funktion*. Wiesbaden: Westdeutscher Verlag.

Maloney, M.P., Ward, M.O., Braucht, C.N. (1985) A revised scale for the measurement of ecological attitudes and knowledge. *American Psychologist*, 30, 787-790.

March, J., & Shapira, Z. (1987). Managerial perspectives on risk and risk taking. *Management Science, 33*, 1401-1418.

Marris, C., Langford, I., & O'Riordan, T. (1996). *Integrating sociological and psychological approaches to public perceptions on environmental risks: Detailed results from a questionnaire survey*. Research Report. Norwich: University of East Anglia.

McCrimmon, K. R., & Wehrung, D. A. (1990). Characteristics of risk taking executives. *Management Science, 36*, 422-435.

McDaniels, T. L., & Gregory, R. S. (1991). A framework for structuring cross-cultural research in risk and decision making. *Journal of Cross-cultural Psychology, 22*, 103-128.

Morgan, M. G. (1993). Risk analysis and management. *Scientific American, 248*, 24-30.

Moscovici, S. (1985). Social influence and conformity. In G. Lindzey & E. Aronson (Eds.), *Handbook of social psychology (Vol.2)* (pp. 347-412). Hillsdale: Erlbaum.

National Research Council (USA) (Ed.) (1990). *Improving risk communication*. Washington: National Academy Press.

Nyland, L. G. (1993). *Risk perception in Brazil and Sweden*. Stockholm School of Economics: Centre for Risk Research.

Opwis, K., & May, R. (1985). *Determinanten der Risikoakzeptanz bei Umweltproblemen*. Forschungsbericht des Psychologischen Instituts Freiburg.

Pilisuk, M., Parks, S. H., & Hawkes, G. (1987). Public perception of technological risk. *The Social Science Journal, 24*, 403-413.

Prester, G., Rohrmann, B., & Schellhammer, E. (1987). Environmental evaluations and participation activities - A social-psychological field study. *Journal of Applied Social Psychology*, *17*, 749-785.

Rayner, S. (1990). *Risk in cultural perspective: Acting under uncertainty*. Klewer: Norwell.

Rayner, S. (1992). Cultural theory and risk analysis. In S. Krimsky & d. Golding (Eds.), *Social theories of risk*. Westport: Praeger.

Renn, O. (1992a). Risk communication: towards a rational discourse with the public. *Journal of Hazardous Materials*, 29, 465- 519.

Renn, O. (1992b). Concepts of Risk: A classification. In S. Krimsky & D. Golding (Eds.), *Social theories of risk* (pp. 53- 82). Westport: Praeger.

Rohrmann, B. (1991). *Risks and benefits of individual activities and living conditions - a cross-cultural comparison* (Research Report). Hamilton/NZ: University of Waikato.

Rohrmann, B. (1994). Risk perception of different societal groups: Australian findings and cross-national comparisons. *Australian Journal of Psychology*, *46*, 150-163.

Rohrmann, B. (1995). *Risk perception research: Review and documentation* (Studies in Risk Communication vol. 48). Juelich: Research Center Juelich. Revision & Update 1999, Studies vol. 69. (*Also published on the WWW; URL* = http://www.kfa-juelich.de/mut/hefte/heft_69.pdf).

Rohrmann, B. (1996). *Perception and evaluation of risks: Findings for New Zealand and cross-cultural comparisons* (Information paper No.55). Canterbury/NZ: Centre for Resource Management, Lincoln University.

Rohrmann, B (1998). The risk notion: Epistemological and empirical considerations. In R. Melchers & M. Stewart, *Integrated Risk Assessment*, 39-45. Rotterdam: Balkema.

Rohrmann, B., & Borcherding, K. (1985). Die Bewertung von Umweltstressoren unter Risiko-Aspekten. In D. Albert (Hg.), *Bericht über den 34. Kongreß der DGfP in Wien 1984* (pp. 851-854). Göttingen: Hogrefe.

Rohrmann, B., & Chen, H. (1999). Risk perception in China and Australia: an exploratory crosscultural study. *Journal of Risk Research 2 (3), 219-241.*

Rowan, K. E. (1991). Goals, obstacles, and strategies in risk communication: A problem-solving approach to improving communication about risks. *Journal of Applied Communication Research*, *November*, 300-329.

Savage, I. (1993). Demographic influences on risk perceptions. *Risk Analysis*, *13*, 413-420.

Schuez, M. (Hrsg.) (1990). *Risiko und Wagnis. Die Herausforderung der industriellen Welt.* Pfullingen: Neske.

Schwarz, M., & Thompson, M. (1990). *Divided we stand.* New York: Harvester Wheatsheaf.

Shapira, Z. (1986). *Risk in managerial decision making.* Jerusalem: Diss/Hebrew University.

Sjöberg, L., & Drottz-Sjöberg, B. (1991). Knowledge and risk perception among nuclear power plant employees. In B. Drottz- Sjöberg (Ed.), *Perception of Risk* (pp. 141-162). Stockholm: Center for Risk Research.

Sjöberg, L. (1998). Worry and risk perception. *Risk Analysis*, 18, 85-94.

Sjöberg, L. (1997). Explaining risk perceptions: An empirical evaluation of cultural theory. *Risk Decision and Policy*, 7, 113-130.

Slovic, P. (1992). Perception of risk: reflections on the psychometric paradigm. In D. Golding & S. Krimsky (Ed.) *Social theories of risk*, pp. 117-523. London: Praeger.

Slovic, P., Fischhoff, B., & Lichtenstein, S. (1980). Facts and fears - understanding risk. In R. C. Schwing & W. A. Albers (Eds.), *Societal risk assessment* (pp. 181-218). New York: Plenum.

Slovic, P., Fischhoff, B., & Lichtenstein, S. (1982). Why study risk perception. *Risk Analysis*, *2*, 83-93.

Slovic, P., Fischhoff, B., & Lichtenstein, S. (1985). Characterizing perceived risk. In R. W. Kates, C. Hohenemser, & J. X. Kasperson (Eds.), *Perilous progress: managing the hazards of technology* (pp. 91-125). Boulder: Westview.

Sokolowska, J., & Tyszka, T. (1995). Perception and acceptance of technological and environmental risks: Why are poor countries less concerned? *Risk Analysis*, *15*, 733-744.

Stern, P. C., & Dietz, T. (1994). The value basis of environmental concern. *Journal of Social Issues*, *50*, 65-84.

Thompson, M., Ellis, W., & Wildavsky, A. (1990). *Cultural theory, or, why all that is permanent is bias*. Boulder: Westview Press.

Tiemann, A. R., & Tiemann, J. J. (1985). Cognitive maps of risk and benefit perceptions. In C. Whipple & V. T. Covello (Eds.), *Risk analysis in the private sector* (pp. 451-468). New York: Plenum Press.

Triandis, H. C. (1989). The self and social behavior in differing cultural contexts. *Psychological Review*, *96*, 506-520.

Tse, D.K., Lee, K., Vertinsky, I., & Wehrung, D.A. (1988). Does culture matter? A cross-cultural study of executives' choice, decisiveness and risk adjustment in international marketing. *Journal of Marketing* 52, 81-95

Tweeddale, H. M. (1994). *Uses and limitations of risk assessment*. Sydney: A Carre.

Vaughan, E. (1995). The significance of socioeconomic and ethnic diversity for the risk communication process. *Risk Analysis*, *15*, 169-180.

Viscusi, W. K., & Magat, W. A. (1987). *Learning about risk: Consumer and worker responses to hazard information*. Cambridge: Harvard University Press.

Wildavsky, A. (1995). *But is it true?*. Cambridge/MA: Harvard University Press.

Winterfeldt, D. v., John, R. S., & Borcherding, K. (1981). Cognitive components of risk ratings. *Risk Analysis*, *1*, 277- 287.

Yates, J. F., & Lee, J. W. (Eds.) (1996). Chinese decision making. In M. H. Bond (Ed.), *Handbook of Chinese psychology* (pp. 338-351). Hong Kong: Oxford University Press.

4

RISK PERCEPTION IN BULGARIA AND ROMANIA

Lennart Sjöberg
Center for Risk Research
Stockholm School of Economics, Stockholm, Sweden

Daniela Kolarova
University of Sofia "St. Kl. Ohridski", Sofia, Bulgaria

Alina-Anca Rucai
Romanian Academy of Sciences, Bucharest, Romania

Marie-Louise Bernström
Center for Risk Research
Stockholm School of Economics, Stockholm, Sweden

TABLE OF CONTENTS

SUMMARY

Risk perception and media risk content were studied in Bulgaria and Romania; results were compared to data obtained in other countries, most notably in Sweden and Brazil. Various groups were sampled for investigating risk perception: nurses, students of economics and engineering, teachers, unskilled workers and homeless people. They rated a large number of general and personal risks, probabilities and consequences of harm, and demand for risk mitigation. It was found that Bulgarian risk perception levels were quite high, on the same level as Brazil and higher than the US data available for comparison. The Romanian data showed considerably lower levels, almost at the level of Scandinavian countries. The differences in levels of perceived risk between Romania and Bulgaria cannot be explained by differences in levels of real risk; Romania was in a clearly more difficult economic situation than Bulgaria. Media contents are therefore suggested as a more potent determinant of perceived risk than real risk. Among detailed results we can also mention that engineering students, followed by MBA students, gave the lowest risk ratings throughout the study. Some homeless people gave, contrary to expectations, low ratings of perceived risk - this could be a function of their limited access to media. Confidence in institutions and media was throughout quite low both in Bulgaria and Romania, but lowest in Bulgaria.

1 INTRODUCTION

Perceived risk creates a social dilemma: the public is often worried about risks that experts deem to be trivial or even totally non-existent. At the same time, experts are worried about other risks that the public tends to ignore. Examples of the first kind are risks of nuclear waste and genetic engineering; the second kind is exemplified by risks of alcohol and dietary habits. The discrepant risk perceptions create social and political tensions. A rapidly expanding field of research is concerned with the description and explanation of risk perception - both of the public and of experts. This work started already in the 1970's, following the seminal work by Sowby (Sowby, 1965) and Starr (Starr, 1969). A second stage of this work is the study of risk communication, a very active field since the middle of the 1980's (Renn, 1992).

Since the mid 80's, there have been several studies which compared risk perception in different countries, including the USA, Hungary (Englander, Farago, Slovic, & Fischhoff, 1986), Poland (Goszczynska, Tyszka, & Slovic, 1991), Norway (Teigen, Brun, & Slovic, 1988), Hong Kong (Keon, 1989), Japan (Kleihesselink & Rosa, 1991), China (Kachelmeier & Shehata, 1992), Sweden and Brazil (Nyland, 1993), see Boholm (1998) for a review. There are at least three reasons why such studies are of interest.

(1) Researchers have been concerned about the generality of basic findings on risk perception. By comparing data from several, sometimes widely different, countries, they have attempted to investigate how general their findings were.

(2) Perceived level of risk for different types of hazards has been assumed to be an important characteristic of the political situation characteristic of a country when it comes to hazard policies. Some of these hazards have, at times, dominated the political scene of a country, such as was the case with nuclear power risks in Sweden in 1979-80.
(3) Studies of culture and psychological phenomena can shed interesting light on the dynamics of such phenomena, and perhaps they can also contribute to the understanding of cultural variation.

The task of comparing whole national populations with regard to risk perception is by no means easy or cheap to carry out. Researchers interested in this field have seldom carried out full surveys of random samples of a whole national population. They have settled for small convenience samples, often only one sample of students. This is obviously unsatisfactory, since the results can be misleading if the sample studied is a quite special one. The only study we have of Soviet risk perception before glasnost is a case in point (Mechitov & Rebrik, 1990). The participants were academy scientists in Moscow and it is well known that experts trained in technology and the natural sciences tend to rate risks of technology to human health and environmental quality as small. This is also what happened in this case, and we can only regret that a broader sampling of Soviet citizens was not carried out. It would have been very interesting to have some data on how ordinary Soviet citizens perceived risk before glasnost.

The demand for fully representative surveys, on the other hand, tends to be stifling. Few researchers have the resources required for such a task. This is not only problematic for economic reasons but also because of the tremendous practical problems that arise in most countries when it comes to drawing random samples of the population, and procuring acceptable response rates for rather complex and exhaustive survey questionnaires.

Furthermore, the focus on representative samples has shadowed a host of other methodological problems that are at least as difficult and important. Indeed, sample representativity is most often conceived of in terms of the distribution of a few key background factors, such as sex, age and education, of the respondents. This is misleading since such background variables usually have only a weak relationship with perceived risk, with the possible exception of gender and general risk ratings (but equal gender distribution is often at least approximately obtained and the variable can always be handled by reporting data for the two genders separately). The background variables, which play an important role in risk perception and related attitudes, are probably more likely to be found in interest and occupational roles. For example, it has been reported that high school academic interest, as revealed in the choice of a study program, was a very potent predictor of perceived risk (Drottz-Sjöberg & Sjöberg, 1991).

In order to overcome some of the above outlined methodological problems, we decided to use a canvassing strategy and to include in our research sample individuals from widely different groups: people working with or studying economics and business, engineering, medical care, and teaching, along with unskilled workers and slum dwellers. The reasons for choosing these particular groups were based on the following assumptions:

(1) Business people, engineers, and students of economics play crucial roles in industrial development, which is in turn responsible for many risks.
(2) Nurses, on the other hand, see the effects of these and other hazards at close hand.
(3) Teachers have an important task since they communicate the values and beliefs of society to children.
(4) Unskilled workers and slum dwellers are the ones that can be assumed to be most directly affected by the risks society creates.

Thus, we conceive of these groups as representing those who are partly responsible for producing risks, those who treat the injuries, those who communicate risk knowledge to the younger generation, and the potential victims. It would of course have been interesting to include also other key groups such as journalists and politicians.

It can be noted that previous work on national comparisons has been concerned mostly with well-educated individuals, perhaps because they were the easiest to reach. Current examples show that the practice of drawing conclusions about national risk perception on the basis of data from only one very special group continues. In a French study, conclusions about the French population were based only on data from a group of graduate students (Karpowicz-Lazreg & Mullet, 1993).

The first time we took the canvassing approach to comparing risk perception on an international scale, was in a study which compared results from corresponding groups[1] in Sweden and Brazil (Nyland, 1993). We found that almost all risks were judged to be more severe by the Brazilian than by the corresponding Swedish groups, while the rank order of risks was very similar, even when quite different groups such as slum dwellers in Sao Paulo and Stockholm engineering students were compared. We also found that students of business and engineering gave lower risk ratings than other groups, which was expected and well in line with previous Swedish results on the correlation between interest and risk (Drottz-Sjöberg & Sjöberg, 1991). The Brazilian slum dwellers gave high risk ratings, also as we expected. An informal media investigation suggested that the Brazilian media (both print and electronic) had much more dramatic and threatening risk content than the Swedish media, especially when it came to domestic risks.

Nyland related these findings to the Cultural Theory of risk perception, as formulated by Douglas and Wildavsky (1982), and argued that perceived risk was not determined by cultural value dimensions, but primarily by media content. However, she had no direct evidence on Cultural Theory variables in her study, only general knowledge to cite from both cultures. The two countries are tremendously different, of course, and if risk perception is mostly determined by culture then one would not expect such large similarities in the ranking of risk as Nyland observed.

The Wildavsky-Dake article in Daedalus in 1990 (Wildavsky & Dake, 1990) brought to our attention the possibility of actually measuring Cultural Theory variables. However, the article was published too late for us to include its scales in the Nyland study. As a result, a new study

[1] Slum dwellers or homeless people were not studied in Sweden because this is a very small group in that country.

was designed and carried out, with Swedish and Brazilian participants who were mainly high school teachers, but some of them taught in elementary school. The study included a subset of 36 risk ratings (out of 100 in the original Nyland study) and translated versions of 53 attitude items provided to us by Dake, that were intended to measure central Cultural Theory constructs, so-called cultural biases. In the present work, we will refer to some of these data when comparing teachers in Bulgaria, Romania, Brazil and Sweden. For a full discussion of risk perception and Cultural Theory, we refer to Sjöberg (1997, 1998). It was found there that risk perception could not be explained, to any noticeable degree, by Cultural Theory.

The momentous changes that started in Eastern Europe at the end of the 1980's, leading to the fall of the totalitarian one-party regimes and the introduction of market economies and free media, set the stage for a corresponding change in risk perception and, perhaps, risk policies. It is probably true that the communist regimes were little concerned about domestic environmental risks. They also seemed to repress information about such risks in the media, which were controlled by the communist party and the governments. In line with this repression of information, the few studies of risk perception that were carried out in Eastern Europe before glasnost and the fall of communism give the impression that people were little worried about domestic risks of technology and environmental degradation. A similar picture is afforded by a recent study from China[2] (Jianguang, 1993).

It is likely that one reason for repressing risk information was that the regimes wanted to avoid creating public pressure for policy changes, which they feared would slow down industrial development. People were suspicious and felt powerless, so why should they accept the risks they believed were exposed to (Sjöberg, 1999 b)? Another reason for repressing risk information was the possible loss of prestige and confidence that such information might have brought about. Perhaps they also judged that environmental degradation was not that serious. All that ended in the former Soviet Union (FSU) with the almost simultaneous occurrence of glasnost and the Chernobyl accident.

It turns out that, when given information that gives cause for alarm, people everywhere are deeply concerned about health, in Eastern Europe just as much as in other parts of the world. Information is the key. This is true even if the economic situation in a country is bad. A comprehensive Gallup study of many countries showed widespread environmental concerns also in the most economically deprived populations (Dunlap, Gallup, & Gallup, 1993).

Two approaches to the explanation of risk perception may be contrasted. The Maslowian model assumes that people ignore all but the most basic needs when they are very poor. Only when they become more affluent do they become more concerned with remote health risks. We have just pointed out that this is a model that does not fit the data very well. For example, Brazilian slum dwellers who certainly were deprived of any trace of affluence and who lived a very dangerous life, were also concerned about the same technology risks as everybody else, if

[2] We know of no such studies from Cuba or North Korea, which otherwise would constitute very interesting countries for a study of risk perception.

not more so. The second model, favored by us, was suggested by Sjöberg and Drottz-Sjöberg (1991) in a study of nuclear power plant personnel. They found that a mild to moderate threat seemed to increase the perceived level of *all* risks, but that a strong threat made other, unrelated, risks seem to be smaller. This suggests an assimilation-contrast model of risk perception. In order to test this model, it is necessary to include groups of subjects that have recently been greatly threatened. In the original study, a group of specialist workers had been threatened by a disastrous explosion. The contrast dynamics could not be observed in the Nyland study, however, perhaps because of the chronic nature of risks associated with living in slums. Presumably, people adapt rather quickly to everyday risks.

Romania and Bulgaria, close in terms of geography and recent history, but widely different in culture and long-term history, were selected for the present study. They are undergoing transitions from one-party communist states where the media were under close government supervision. When the study was conducted, four years had passed since the fall of communism. The information flow was certainly larger and much freer than before, in both countries. However, no data seemed to exist on the details and quantitative characteristics of changes in media content. Moreover, no studies known to us had been carried out on risk perception in either of these countries.

Earlier work in our unit was available, which made it possible to compare risk perception in Bulgaria and Romania with Sweden and Brazil, two countries that are widely different from one another, and also from Bulgaria and Romania.

The dimensions that are relevant in comparing perceived risk levels in different countries need to be carefully considered. We can here only give an informal first attempt toward such a discussion. However, some dimensions seem to be obvious:

(1) True, objective, levels of risk as reflected in expected length of life and causes of death. The higher the real risk, the more should people be concerned.
(2) Mass media coverage of risks. With more mass media coverage of risks people should show more concern.
(3) Economic condition of the society and standard of living. If people are poor they might either show more concern about all risks, or they might be mostly concerned with everyday survival matters, and neglect more long-term risks.
(4) Social protection policies, for example medical care. With a better social security net people might be less concerned about risks.
(5) Changes. Rapid and profound societal changes create anxiety and should therefore have the effect of increasing perceived risk levels.
(6) The size of a country and its population have been suggested to be important factors in cross-cultural comparisons of risk perception level (Goszczynska, Tyszka, & Slovic, 1991), due to the availability heuristic (Tversky & Kahneman, 1973). A large country with a high population density provides many more instances of accidents, which are likely to be stored in people's memory and influence their judgments.

We compare Sweden, Brazil, Romania and Bulgaria. Sweden is an affluent welfare state with a long life expectancy. Swedish society emphasizes security and has a well functioning "security net" for all citizens. The level of crime, especially violent crime, is low and accidents are relatively rare[3]. The Estonia disaster in September 1994 with about 900 victims was the worst disaster since the famine of 1868-1870 (Nelson, 1988)[4] and the war events of 1718[5]. The climatic and geographical conditions of Sweden further contribute to stability and a low level of objective risk (no earthquakes, volcanoes, hurricanes or floods). Nature is benign: there are virtually no dangerous animals or poisonous plants. A very long history of peace (since the beginning of the 19th century) and, until quite recently, an ethnically very homogenous society have further contributed to a low level of risk caused by international or social conflicts. Since 1921, Sweden is a fully democratic society and used to having a situation with a stable majority of Social Democrats in the public arena. However, since the 1970's the political situation has become much more unstable[6]. Overall, its economic development has been positive, but economic growth has shown a tendency to slow down. Workers' real net wages have not changed since the beginning of the 1970's, although households have a considerably higher standard of living now due to the fact that most women also are gainfully employed.

Brazil is a society considerably more filled with dangers of violent crimes and various types of disastrous accidents. Nyland gives a fuller comparative discussion of Sweden and Brazil (Nyland, 1993). Brazil's economy is growing rapidly, but its wealth is distributed in a very uneven manner.

At the time of the study in 1993, Bulgaria and Romania were in transition from communism to a market economy and democracy. The situation was similar in the two countries - economically and politically turbulent with a decreasing standard of living, but the economic situation was worse in Romania. It is also noteworthy that the media were quite outspoken about risks, specially about domestic risks. They provided much more information than during the preceding communist period, but the media risk contents was more extensive in Bulgaria. (Media contents were analyzed in a separate study by Sjöberg et al., 1996). The purpose of the present study was to investigate, using a canvassing strategy, the present[7] level of perceived risk in Romania and Bulgaria, and to compare it with corresponding data from Brazil and Sweden.

---

[3] It should be mentioned, however, that road traffic fatalities are about 600 per year, suicides around 2000.

[4] Many persons died, unclear how many directly due to famine.

[5] In that year, about 3000 soldiers, leaving a war in Norway on a mountain route, froze to death when they met with unexpected and drastic wheather changes.

[6] This is largely due to a constitutional reform in 1970 which made Parliamentary representation of a party strictly proportional to its share of votes, provided that it exceeded 4%. As a further contributory factor, there is a slowly decreasing lack of confidence in politicians and a tendency for voters to opt for small parties with specialized agenda, such as the "Green" party (represented in Parliament in 1988-91, and again since 1994) or the Christian Democrats (first represented in 1991).

[7] Unfortunately no data on risk perception in communist Bulgaria and Romania could be located.

1.1 Questionnaires

Two questionnaires of the same type were administered in Sofia, Bulgaria and Bucharest, Romania. The translations from English to Bulgarian and Romanian and back into English were carried out by professional translators. The Bulgarian survey contained 327 questions and in the Romanian questionnaire the respondents were asked 322 questions. In both countries there were questions referring to background data. The respondents were asked to specify their age, sex, civil status, educational level, and profession. The remaining questions were divided in five different sections. The first section covered 100 hazards deriving from different human activities, chemical substances, or technologies. The subjects rated the risk of death stemming from each one of these hazards for a period of about five years. The risk was specified to society as a whole, and not dependent or limited to people actually exposed to the hazard in question. There was a scale ranging from 0 = "non-existent risk" to 100 = "extremely large risk", in order to help the subjects to make their ratings. In addition, the subjects in both countries were asked to rate country-specific risks. In Bulgaria, 13 additional risk questions were added, and 14 questions were added in the Romanian survey.

The second section was divided into six parts. The first part treated personal risks ("How large do you judge the risk to be for you personally of the following?"), and 28 questions were posed. The respondent was asked to rate these risks on a 0 to 6 scale, ranging from "non-existing risk" to "extremely large risk". The questions were the same in both countries, and the same 28 risks were used in the other parts, two to six. The risks were:

1. To be injured by smoking
2. To be injured by alcohol consumption
3. To be injured by exhausts from motor vehicles
4. To be injured by industrial pollution
5. To be injured by climate changes/ green house effects
6. To be assaulted
7. To be injured by inadequate food diet
8. To have a serious car accident
9. To be struck by lightning
10. To be injured by the depletion of the ozone layer
11. To be injured by domestic nuclear power
12. To be injured by foreign nuclear power
13. To be injured by natural background radiation
14. To be injured by unsuitable food additives
15. To be lonely
16. To be injured in a train accident
17. To be unemployed
18. To get AIDS
19. Not afford adequate housing
20. Not afford sufficient nutritious and tasty food
21. Not have an adequate treatment when being ill
22. To have water storage problems
23. To be affected by dirty public places
24. To be affected by national social conflicts
25. To be affected by war or other conflicts in the nearby region
26. Not be able to get a good education for yourself and your family
27. To be poor
28. To be attacked by wild dogs in the street

The second part dealt with questions concerning general risks. The following question was posed to the subjects: "How large do you judge the risk to be for people generally in the country?". The same 28 risks were used, and rated on a 0 to 6 scale ranging from "non-existing risk" to "extremely large risk".

The third part dealt with the demand for risk mitigation (cp. Sjöberg, 1999 a), and the subjects were asked to what extent they thought the risks can be tolerated or should be reduced. A 0 to 6 scale was used for this purpose ranging from "the risk could absolutely be tolerated" to "the risk should be very strongly reduced or completely eliminated". In the fourth part of the survey, the respondents were asked to judge to what extent they could protect themselves against the 28 risks. The scale used for this purpose was 0 "non-existing possibility" to 6 "enormously large possibility". The fifth part dealt with the probabilities that a specific risk would harm the respondent personally or generally. Some of the respondents were asked if they would be harmed personally, and the rest were asked if the risk would cause harm, generally. The subjects were asked to judge these probabilities on a 0 to 1 scale where 0 equals "harm will certainly not occur" and 1 "very large harm is certain to occur". In the sixth part of the survey the respondents were asked to estimate how severe the consequences of harm would be for several different risks. This part also had two alternative formulations, similar to the question in part five. The scale used in this part was from 0 = "no severity at all" to 6 = "extremely severe".

The third section of the questionnaires treated the topic of politics and the possibilities that people have to influence the policies in their countries. Fourteen topics were covered: taxes, education, nuclear power, immigrants, defense, environment, cost of housing, food prices, crime, unemployment, price of electricity, external policy, agricultural policy, and consumption of goods. The scale used was from 0 = "enormously large possibility" to 6 = "enormously small possibility". The fourth section was related to the topic of anxiety ("How often have you experienced the following during the last five years?"), and 12 different questions were posed on that subject:

1. Have you had nightmares?
2. Has it happened that you have seen a doctor without the doctor finding anything wrong with you?
3. Have you had difficulty to keep your thoughts under control due to worries?
4. Have you had difficulties to sleep?
5. Do you keep thinking about the possibility that you or close family can be struck by illness or accidents?
6. Do you feel nervous when flying?
7. Do you feel it is dangerous to be outdoors late at night?
8. Have you avoided to take a medicine prescribed by a doctor because you were afraid of side effects?
9. Does it make you nervous when you are a passenger in a car and the road is slippery, if it is dark and rainy, or otherwise difficult to drive?
10. Do you think about the possibility of fire in a house where you are?
11. Is a violent thunderstorm worrisome for you?
12. Do you worry about burglary in your home?

In this part there was a 1 to 5 scale where 1 equals "never" and 5 "extremely often". In the fifth section, the subjects were asked to what extent they had confidence in risk related information from 15 different sources. The sources were: State television, other TV channels, morning paper that I read, evening paper that I read, state radio, other domestic radio (only Bulgaria), foreign radio, friends, doctors, teachers, priests, central government, local government, labor union, and experts. In this case the scale was from 1 = "none at all" to 5 = "to a very high degree".

1.2 Subjects

In Bulgaria a total of 240 persons participated in the study, in Romania 192. The subjects formed six different samples, and the categories were the following: Nurses, workers, graduate students of engineering, students of economics, teachers, and slum dwellers. See Table 1.

Table 1: **THE NUMBER OF RESPONDENTS PARTICIPATING IN THE STUDY**

	Nurses	Workers	Students of engineering	Students of economics	Teachers	Slum dwellers	Total
Bulgaria	40	40	40	40	40	40	240
Romania	30	34	35	34	29	30	192
Total	70	74	75	74	69	70	432

Since the sizes of the survey samples that we could obtain from these two countries are rather small, large mean differences will be required to establish any statistical significance. Country will be the most important independent variable in the present study.

2 RESULTS

2.1 Comparisons between Bulgaria and Sweden

The results obtained from the Bulgarian groups were analyzed separately and then compared with the Swedish groups. Means and standard deviations were computed for each of the 100 risks derived from different human activities, substances and technologies. The Bulgarians perceived these 100 general risks as larger than the Swedes did. The Bulgarian respondents viewed nuclear risks as particularly important hazards. Among the 10 risks that obtained the highest risk perception ratings were nuclear risks, warfare, heroin and narcotics, AIDS, and ozone depletion. The 10 hazards that were perceived as lowest were the following: hair dyes, fluorescent lamps, sun bathing, cooking stoves, solar electric power, bicycles, hydroelectric power, aspirin, power mowers, saccharin and football. The 13 extra risks that were introduced in the Bulgarian questionnaire obtained quite high risk perception ratings and were ranked among the upper half of all risks. Table 2 presents the overall means and standard deviations for each group in Bulgaria and in Sweden. The over-all difference in means between the two countries was statistically significant ($p < 0.0005$).

Table 2: **COMPARISON OF THE OVERALL MEANS AND STANDARD DEVIATIONS FOR EACH GROUP IN BULGARIA AND SWEDEN**

	All respondents	Nurses	Workers	Students of engineering	Students of economics	Teachers	Slum dwellers
Bulgaria							
Mean	49.78	56.83	52.77	42.37	45.20	53.88	47.64
SD	14.04	14.47	9.95	15.78	11.00	15.35	11.56
N	240	40	40	40	40	40	40
Sweden							
Mean	31.61	40.28	29.84	26.94	29.31	-	-
SD	14.01	15.69	8.20	13.57	14.13	-	-
N	119	30	30	30	29	-	-
Diff.	-18.17	-16.55	-22.93	-15.43	-15.89	-	-
Prob.	0.0005	0.0005	0.0005	0.0005	0.0005	-	-

The highest levels of perceived risk in Bulgaria were found among nurses, followed by workers and teachers. The students of engineering perceived these risks as lower than the other groups. A one way analysis of variance showed that there were also statistically significant differences among the groups in Bulgaria ($F = 7.15$, $p < 0.001$). Similar results were found in Sweden (Nyland, 1993, pp. 55-56). A two way analysis of variance, using the independent variables

Table 3: **MAXIMAL MEAN DIFFERENCES BETWEEN COUNTRIES FOR THE 100 FIRST RISK QUESTIONS. THE LARGEST DIFFERENCES BETWEEN THE COUNTRIES**

Risks	Bulgarian mean	Swedish mean	Difference
1. Diagnostic X-rays	52.09	15.10	36.99
2. Nitrogen fertilizers	63.55	27.20	36.35
3. Antibiotics	52.23	15.10	35.93
4. Food irradiation	65.70	30.50	35.20
5. Nuclear power (routine operation radiation)	74.11	39.00	35.11
6. Nuclear power (production)	69.54	34.50	35.04
7. Caffeine	48.21	14.30	33.91
8. Mushroom hunting	41.50	7.90	33.60
9. Crime	76.39	47.50	28.89
10. Pregnancy and child birth	39.62	11.20	28.42

Table 4: **MINIMAL MEAN DIFFERENCES BETWEEN COUNTRIES FOR THE 100 FIRST RISK QUESTIONS**

Risks	Bulgarian mean	Swedish mean	Difference
1. Heroin	78.52	77.00	1.52
2. DDT	54.64	52.20	2.44
3. Asbestos	51.78	54.50	-2.72
4. Nuclear weapons (war)	82.11	79.20	2.91
5. Sun bathing	25.04	22.00	3.04
6. Bicycles	22.94	19.70	3.24
7. Alcohol related accidents	63.00	59.70	3.30
8. Nerve gas	65.15	69.50	-4.35
9. Motorcycles	34.77	41.50	-6.73
10. 2,4,5, T Dioxins	61.26	53.50	7.76

country and group, showed that there were significant differences among all the studied groups, across both countries ($F = 24.86$, $p < 0.001$). The two-way interaction of the variables country and group was not found to be significant ($F = 1.24$, ns).

The tables below (3 and 4) show which risk perception questions gave rise to the largest and the smallest differences between the two countries. These maximal mean differences were due to overall higher Bulgarian ratings in relation to the Swedish risk data. Many of these large mean differences could be associated with some kind of health aspect or environmental hazard (see Table 3).

The minimal differences between Sweden and Bulgaria occurred among different kinds of risks, technological as well as health related (see Table 4).

Figure 1 shows how the Bulgarians rated the 100 risks compared to the Swedish data. The figure above shows that there was a high correlation between the risk ratings for the 100 studied risks. Both in Sweden and Bulgaria these risks were ranked in much the same way, but the Bulgarian respondents rated the risks much higher than the Swedes did.

Figure 1: **MEAN RATINGS OF 100 RISKS BY THE SWEDISH AND BULGARIAN SAMPLES**

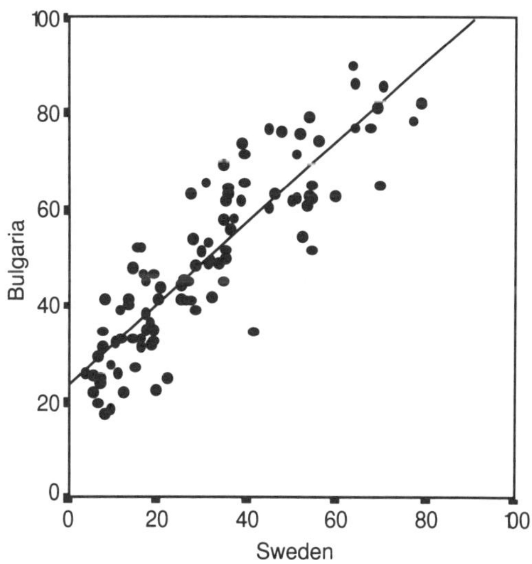

2.2 Comparison between Romania and Bulgaria

The Romanians rated the 100 general risks as smaller than the Bulgarians did. The Romanian as well as the Bulgarian samples viewed the nuclear risks as particularly important hazards. Among the 10 risks that obtained the highest risk perception ratings in Romania were nuclear risks, AIDS, crime, handguns, and narcotics. The 10 hazards that were perceived as lowest were the following: hair dyes, fluorescent lamps, satellites, large constructions, bicycles, heavy farming machines, microwave ovens, aspirin, power mowers and football. There were large similarities with the results obtained for the Bulgarian respondents. The 14 extra risks that were introduced in the Romanian questionnaire obtained quite high risk perception ratings and were ranked among the upper half of all risks. Table 5 presents the overall means and standard deviations for each group in Bulgaria and in Romania. The difference in means between the two countries was statistically significant ($p < 0.0005$).

Table 5: **COMPARISON OF THE OVERALL MEANS AND STANDARD DEVIATIONS FOR EACH GROUP IN BULGARIA AND ROMANIA**

Bulgaria							
	All respondents	Nurses	Workers	Students of engineering	Students of economics	Teachers	Slum dwellers
Mean	49.78	56.83	52.77	42.37	45.20	53.88	47.64
SD	14.04	14.47	9.95	15.78	11.00	15.35	11.56
N	240	40	40	40	40	40	40
Romania							
Mean	39.20	51.01	35.91	34.78	39.27	34.24	40.73
SD	15.79	15.40	16.33	14.16	14.93	15.50	13.11
N	190	30	34	34	35	28	29
Diff.	-10.58	-5.82	-16.87	-7.58	-5.93	-19.64	-6.90
Prob.	0.0005	ns	0.0005	0.03	ns	0.0005	0.03

The highest levels of perceived risk in Romania were found among nurses, followed by slum dwellers and students of economics. The teachers perceived these risks as lower than the other groups. A one way analysis of variance showed that there were also statistically significant differences among the groups in Romania ($F = 5.34$, $p < 0.001$), to be compared with results from the comparison between Sweden and Bulgaria. A two way analysis of variance, using the independent variables country and group, showed that there were significant differences among all the studied groups, across both countries ($F = 11.16$, $p < 0.001$). The interaction effects between the two independent variables were significant ($F= 3.32$, $p< 0.006$) (see Figure 2). Figure 2 displays how the different groups in both countries rated the first 100 risks. The interaction effects present in the data can be studied in this figure.

Figure 2. **GRAPHIC REPRESENTATION OF THE GENERAL MEANS FOR EVERY STUDIED GROUP. THE SCALE OF THIS QUESTION GOES FROM 0 TO 100.**

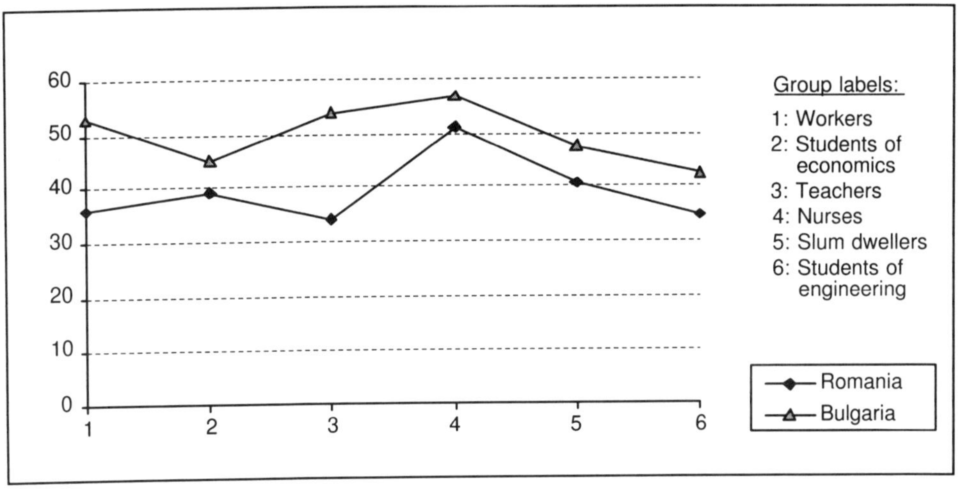

An interesting question related to the above figure is if gender could explain a part of the mean differences among the groups. In a three-way ANOVA-test with group, country and gender as independent variables, the following results were obtained: the variables country ($F = 59.58$, $p < 0.0005$) and gender ($F = 14.21$, $p < 0.0005$) were significant variables in explaining the mean differences, but the independent variable group did not any longer explain any variation in mean ratings ($F = 0.89$, ns). The two way interactions were all significant. These interactions were the following: country and group ($F = 3.74$, $p < 0.001$), country and gender ($F = 4.05$, $p < 0.02$), and gender and group ($F = 3.31$, $p < 0.01$).

When we examine the gender effects in the collected data for each country separately, the following results were obtained by two-way ANOVA tests using gender and group as independent variables. In Bulgaria, the mean differences for the first 100 risks could be explained by the variable group ($F = 6.61$, $p < 0.0005$), but not by gender ($F = 0.05$, ns). In Romania, the results were reversed, which means that gender explained the mean differences ($F = 17.8$, $p < 0.0005$) but not the group ($F = 1.18$, ns). There were interaction effects in the Romanian data ($F = 2.90$, $p < 0.02$) that were not found in Bulgaria ($F = 1.02$, ns). These results could also be interesting to study in a graphic representation (see Figures 3 and 4).

Figure 3 shows that the mean differences between men and women in Bulgaria were not large. Gender did not seem to affect how the respondent rated these 100 risks. As mentioned earlier, there were no interactions between the two variables in Bulgaria. The similarity in mean ratings between men and women is quite intriguing. Gender differences in risk perception are quite common.

Figure 4 shows that there were rather large differences between men and women in Romania. Women generally rated these risks as larger than men did. The only exception was the category

Figure 3. **A GRAPHIC REPRESENTATION OF THE MEAN RATINGS BY GENDER AND BY GROUP IN BULGARIA.**

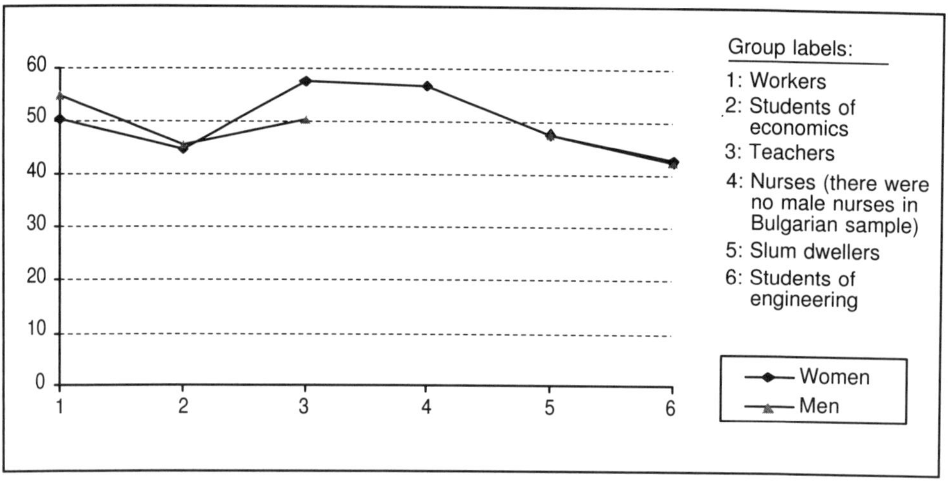

workers in which men rated the risks as larger than women did. Figure 4 shows also that there are interactions between the variables gender and group.

In Figure 5 the Bulgarian groups' mean ratings are compared to the Romanian men and women. The following conclusion can be drawn: the Bulgarian respondents were more close to the Romanian women in their ratings of the first 100 risks than the Romanian men. The ratings of the slum dwellers (group 5) and students of engineering (group 6) were especially close (see

Figure 4. **A GRAPHIC REPRESENTATION OF THE MEAN RATINGS BY GENDER AND BY GROUP IN ROMANIA.**

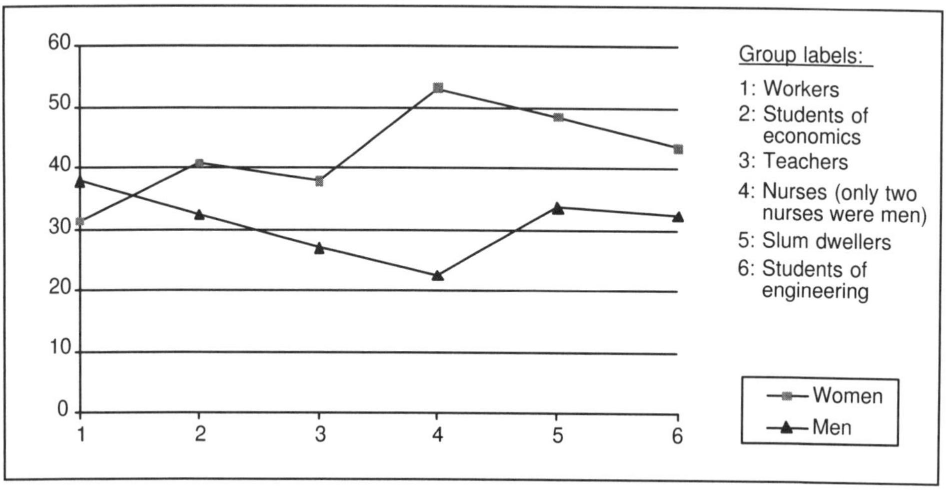

Figure 5. **A GRAPHIC REPRESENTATION OF THE MEAN RATINGS BY GENDER AND BY GROUP IN ROMANIA AND ONLY BY GROUP IN BULGARIA.**

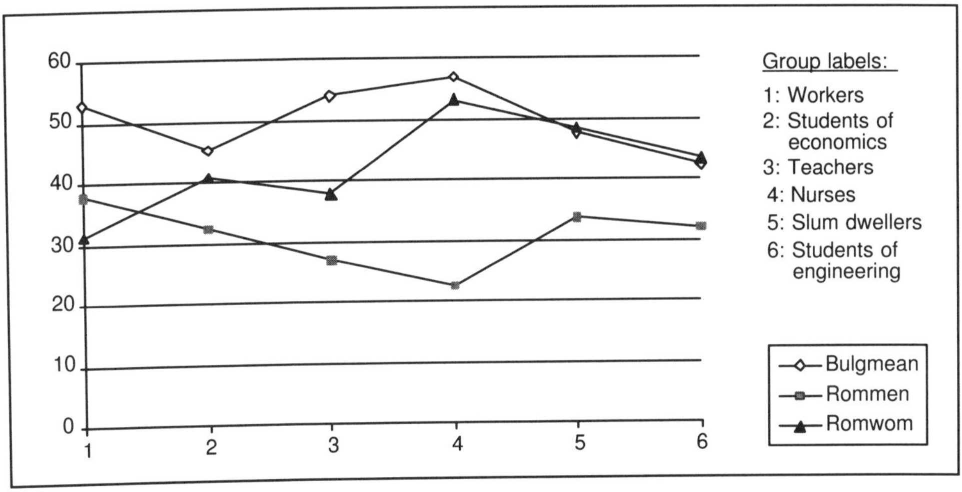

Figure 6. **A GRAPHIC REPRESENTATION OF WOMEN'S MEAN RATINGS BY GROUP IN ROMANIA AND BULGARIA.**

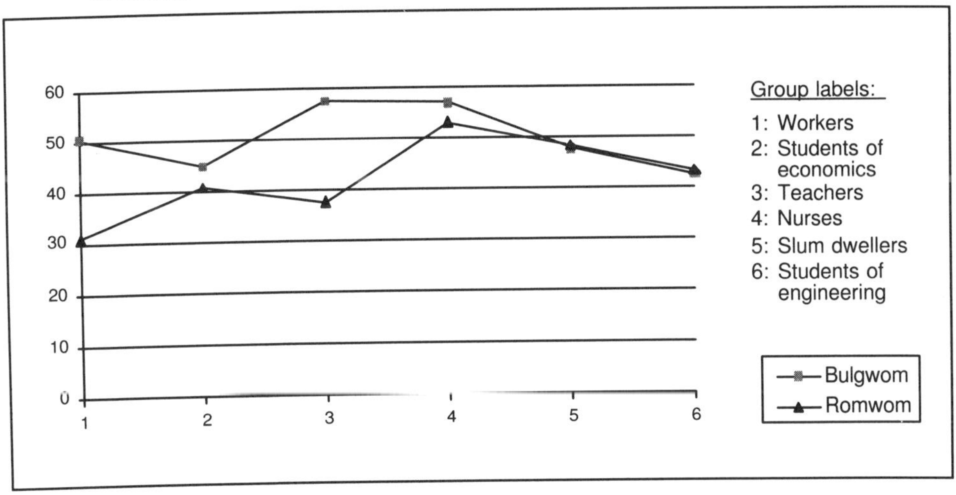

Figure 5). A comparison of the women's ratings in both countries is presented graphically in Figure 6.

The mean ratings of the women were quite similar in both countries. The two exceptions in this case were workers and teachers. A more general conclusion in relation to Figures 3, 5 and 6 is that the group that seemed to diverge the most from the others are the Bulgarian men. One possible explanation would be that men in Bulgaria were older than the women in Bulgaria, and also older than the Romanian respondents. Older people tend to rate risks as larger than young persons do, and there could therefore be age differences in the Bulgarian data that would explain why Bulgarian men rated these risks in the same way as the Bulgarian women did. A t-test of the independent samples women and men against the dependent variable age showed that men in

Table 6: **MAXIMAL MEAN DIFFERENCES BETWEEN COUNTRIES FOR THE 100 FIRST RISK QUESTIONS**

Risks	Bulgarian mean	Romanian mean	Difference
1. Lead paint	64.91	21.85	43.06
2. Leaded gasoline	71.47	37.69	33.78
3. Non-nuclear electric power	35.47	67.26	-31.79
4. Asbestos	51.78	20.32	31.46
5. Antibiotics	52.23	21.05	31.18
6. Oral contraceptives	46.02	17.53	28.49
7. Service in the Army	44.54	16.72	27.82
8. Auto exhaust fumes	61.78	34.62	27.16
9. Mercury	75.84	50.12	25.72
10. Nitrogen fertilizers	63.55	38.14	25.41

Table 7: **MINIMAL MEAN DIFFERENCES BETWEEN COUNTRIES FOR THE 100 FIRST RISK QUESTIONS**

Risks	Bulgarian mean	Romanian mean	Difference
1. Skyscraper fires	63.43	63.12	0.31
2. National defense	48.98	50.14	-1.16
3. Saccharin	18.79	17.56	1.23
4. Crime	76.39	74.81	1.58
5. Handguns	71.87	70.14	1.73
6. Dynamite	63.35	65.19	-1.84
7. Nuclear power (routine operation radiation)	74.11	72.2	1.91
8. Acquired Immune Deficiency (AIDS)	77.22	79.30	-2.08
9. Skyscrapers (construction and maintenance)	41.34	38.85	2.49
10. Food preservatives	31.88	34.85	-2.97

Bulgaria were older than the women (diff. = 2.47, $p < 0,02$). In a three-way-ANOVA-test for the Bulgarian data with group, age and gender as independent variables and mean ratings as the dependent variable, the following results were obtained: the variable group ($F = 6.22$, $p < 0.0005$) was significant in explaining the mean differences, but the independent variables gender ($F = 0.05$, ns) and age ($F = 0.08$, ns) did not explain any variation in the Bulgarian mean ratings ($F = 0.89$, ns). The two way interactions were all insignificant. Hence, age did not explain the findings. Tables 6 and 7 below show which risk perception questions gave rise to the largest and the smallest differences between the two studied countries. These maximal mean differences were due to the overall higher Bulgarian ratings in relation to the Romanian risk data. The only exception was non-nuclear electric power. Many of these large mean differences could be associated to some kind of health aspect or environment hazard. The minimal differences between Romania and Bulgaria occurred among different kinds of risks, technological as well as health related.

Figure 7. **MEAN RATINGS OF 100 RISKS BY THE BULGARIAN AND ROMANIAN SAMPLES.**

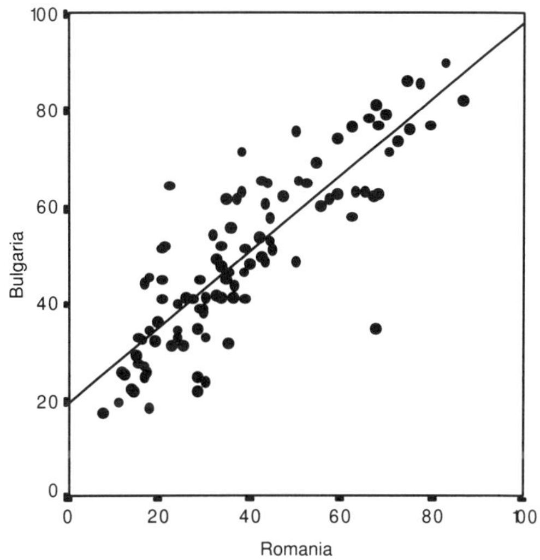

Figure 7 shows that there was a large correlation between the mean risk ratings for the 100 studied risks in Bulgaria and Romania. These risks were ranked in approximately the same way in Romania and Bulgaria, but the Bulgarian respondents rated the risks as larger than the Romanian samples did. The only exception was non-nuclear electric power, a risk that the Romanian respondents rated as larger than the Bulgarians did.

2.3 Comparisons between Sweden and Romania

Table 8 presents the overall means and standard deviations for each group in Sweden and in Romania. The difference in means between the two countries was statistically significant ($p < 0.0005$), which implies that the Romanians perceived these hazards as larger than the Swedish respondents did.

A one way analysis of variance showed that there were also statistically significant differences among the groups in Romania and Sweden. The two ANOVA tests showed that the differences of means were significant ($F= 8.82$ and $p < 0.0005$). The two way interaction between the independent variables was not significant ($F = 0.34$, ns).

Figure 8 shows that there was a high correlation between the mean risk ratings for the 100 studied risks. These risks were ranked in approximately the same way in Romania and Sweden, but the Romanian samples rated these risks as somewhat larger, on the average, than the Swedes did.

Table 8: **COMPARISON OF THE OVERALL MEANS AND STANDARD DEVIATIONS FOR EACH GROUP IN ROMANIA AND SWEDEN**

Romania							
	All respondents	Nurses	Workers	Students of engineering	Students of economics	Teachers	Slum dwellers
Mean	39.20	51.01	35.91	34.78	39.27	34.24	40.73
SD	15.79	15.40	16.33	14.16	14.93	15.50	13.11
N	190	30	34	34	35	28	29
Sweden							
Mean	31.61	40.28	29.84	26.94	29.31	-	-
SD	14.01	15.69	8.20	13.57	14.13	-	-
N	119	30	30	30	29	-	-
Diff.	7.59	10.73	6.07	7.84	10.73	-	-
Prob.	0.005	0.01	ns	0.03	0.01	-	-

Figure 8. **MEAN RATINGS OF 100 RISKS BY THE SWEDISH AND ROMANIAN SAMPLES.**

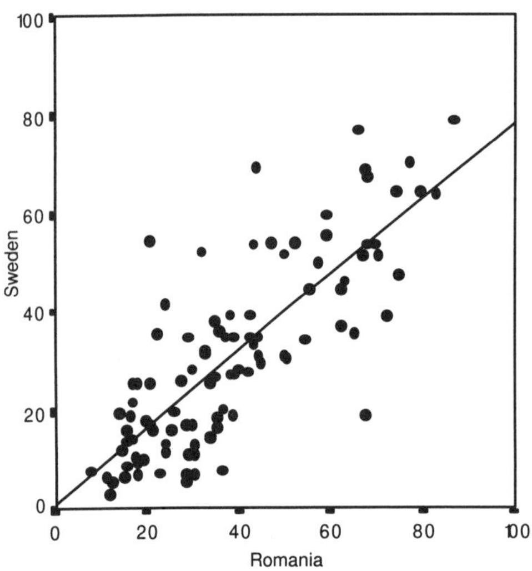

Figure 9. **MEAN RATINGS OF PERSONAL RISKS BY THE BULGARIAN AND ROMANIAN SAMPLES.**

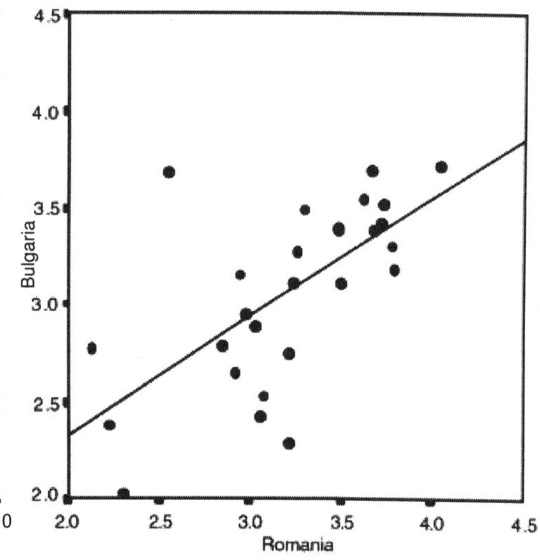

2.4 Personal risks

Table 9 below shows the mean magnitude for the studied perceived personal risks for all groups both in Bulgaria and Romania. There were no statistically significant differences between the corresponding groups, in the two countries, as it is shown in Table 10. A one way analysis of variance showed that there were statistically significant differences among the studied groups in

Table 9: **MEANS FOR THE PERSONAL RISK QUESTIONS**

	All respondents	Nurses	Workers	Students of engineering	Students of economics	Teachers	Slum dwellers
Bulgaria							
Mean	3.10	3.58	3.16	2.87	2.91	3.23	2.86
SD	1.13	1.18	1.23	0.85	1,02	1.31	0.99
N	239	40	40	40	40	39	40
Romania							
Mean	3.22	3.43	3.66	2.89	3.02	2.98	3.35
SD	1.29	1.20	1.37	1.12	1.17	1.42	1.35
N	192	30	34	35	34	29	30
Diff.	0.12	-0.15	0.50	0.03	0.11	-0.25	0.49
Prob.	ns	ns	ns	ns	ns	ns	ns

Bulgaria ($F = 2.60$, $p < 0.03$). A corresponding test was performed on the Romanian data, but the differences among the groups were not significant, ($F = 1.83$, ns).

Nurses in Bulgaria and workers in Romania gave the highest ratings of personal risk. The lowest ratings were given by slum dwellers in Bulgaria and by students of engineering in Romania. A two way analysis of variance, with country and category as independent variables, showed significant differences between the personal risk perceptions ($F = 2.10$, $p < 0.02$). The interaction between the two independent variables was not significant, ($F = 1.24$, ns).

The ratings of personal risks in Bulgaria and Romania were similar, but there were three exceptions: the risks to be assaulted, to be struck by lightning and to be attacked by wild dogs in the street. This can also be seen in Table 10 below:

Table 10: **MAXIMAL CROSS-NATIONAL DIFFERENCES IN PERCEIVED PERSONAL RISKS**

Risks	Bulgarian mean	Romanian mean	Difference
1. To be assaulted	3.69	2.55	1.14
2. To be struck by lightning	2.29	3.21	-0.92
3. To be attacked by wild dogs in the street	2.76	2.12	0.64
4. To get AIDS	2.43	3.05	-0.62
5. To have a serious car accident	3.19	3.78	-0.59

The risks in Table 11 are the ones with the smallest cross-national differences. These tables (10 and 11) show that the differences between the two countries were not especially important. The respondents in both countries perceived these personal risks in the same way.

Table 11: **MINIMAL CROSS-NATIONAL DIFFERENCES IN PERCEIVED PERSONAL RISKS**

Risks	Bulgarian mean	Romanian mean	Difference
1. Not afford sufficient nutritious and tasty food	3.27	3.25	0.02
2. To be injured by inadequate food diet	2.95	2.98	-0.03
3. Not have a good treatment when being ill	3.70	3.67	0.03
4. To be injured by industrial pollution	3.55	3.61	-0.06
5. To be injured in a train accident	2.78	2.85	-0.07

2.5 General risks

Table 12 below shows the mean magnitude for the studied perceived general risks for all groups both in Bulgaria and Romania. Differences in means between the two countries were not statistically significant, except for workers ($p < 0.03$) and slum dwellers ($p = 0.01$) as it is shown in Table 12. A one way analysis of variance showed that the group differences in Bulgaria were significant ($F = 3.01$, $p < 0.01$), as well as among the groups in Romania ($F = 4.11$, $p < 0.001$). The results for all groups in Romania and Bulgaria show that the general risks were rated higher than the personal risks in all groups in both countries.

Table 12: **MEANS FOR THE GENERAL RISKS IN BULGARIA AND ROMANIA**

Bulgaria							
	All respondents	Nurses	Workers	Students of engineering	Students of economics	Teachers	Slum dwellers
Mean	3.65	4.08	3.46	3.27	3.69	3.80	3.58
SD	1.04	0.98	1.21	0.99	0.81	1.20	0.87
N	238	40	40	39	40	39	40
Romania							
Mean	3.84	4.08	4.09	3.21	3.71	3.76	4.31
SD	1.10	0.98	1.16	0.91	0.92	1.06	1.25
N	190	29	34	35	34	29	29
Diff.	0.19	0.00	0.64	-0.06	0.02	-0.05	0.73
Prob.	ns	ns	0.024	ns	ns	ns	0.01

The highest means for the general risks were obtained in the group of Bulgarian nurses and the group of Romanian workers. The lowest means were noted in the group of engineers in both countries. A two way analysis of interactions with the independent variables country and category revealed significant differences among the groups and the countries in relation to general risk perception ($F = 3.60$, $p < 0.001$). The interaction was not significant ($F = 2.20$, ns).

Figure 10 also shows the correlation between the countries, but again the same three risk ratings differ: the risks to be assaulted, to be struck by lightning and to be attacked by wild dogs in the street. Two of them are also among the five maximal cross-national differences in Table 13; minimal differences are given in Table 14.

Figure 10. **MEAN RATINGS OF GENERAL RISKS BY THE BULGARIAN AND ROMANIAN SAMPLES.**

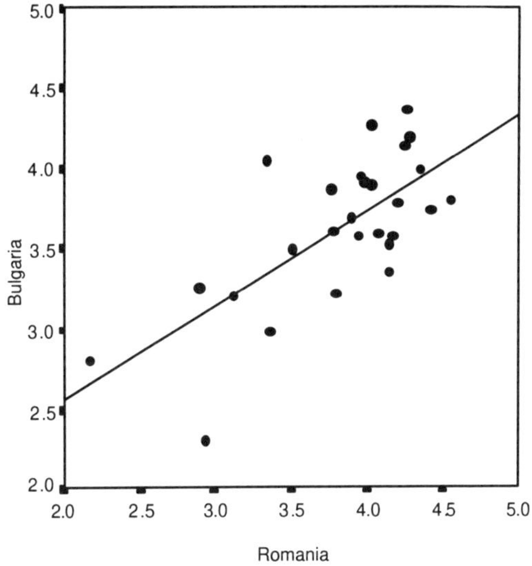

Table 13: **MAXIMAL CROSS-NATIONAL DIFFERNCES IN PERCEIVED GENERAL RISKS (BULGARIA AND ROMANIA WERE COMPARED)**

Risks	Bulgarian mean	Romanian mean	Difference
1. To be affected by national social conflicts	3.36	4.14	-0.78
2. To be injured by foreign nuclear power	3.81	4.54	-0.73
3. To be assaulted	4.05	3.33	0.72
4. To get AIDS	3.75	4.42	-0.67
5. To be attacked by wild dogs in the street	2.81	2.16	0.65

It could be of interest to note that the risks reaching the maximal and minimal mean differences for the personal as well as for the general risks for both countries were almost the same (see also Tables 10 and 11). Three out of five maximal differences between means were the same for perceived personal risks and for the perceived general risks. Among the minimal mean differences, four risks out of five were the same in both cases.

Table 14: **MINIMAL CROSS-NATIONAL DIFFERENCES IN PERCEIVED GENERAL RISKS**

Risks	Bulgarian mean	Romanian mean	Difference
1. To be injured by inadequate food diet	3.50	3.50	0.00
2. To be injured by smoking	3.95	3.96	-0.01
3. Not afford sufficient nutritious and tasty food	3.92	3.98	-0.06
3. Not have a good treatment when being ill	4.21	4.27	-0.06
5. To be injured in a train accident	3.21	3.12	0.09
5. To be poor	4.15	4.24	-0.09

2.6 Demand for risk mitigation and control over the risks

Table 15 below contains the results related to the questions of risk mitigation and perceived control of different types of hazards. Differences in means between the two countries were not statistically significant, except for all respondents together ($p = 0.005$) and slum dwellers ($p < 0.001$), as it is shown in Table 1. A one way analysis of variance shows that the differences between the categories were significant in Bulgaria ($F = 2.30$, $p < 0.05$), but not significant among the groups in Romania ($F = 0.94$, ns).

Table 15: **MEANS FOR DEMAND OF RISK MITIGATION AND CONTROL OF THE HAZARDS IN BULGARIA AND IN ROMANIA**

Bulgaria							
	All respon-dents	Nurses	Workers	Students of engineering	Students of economics	Teachers	Slum dwellers
Mean	4.08	4.48	3.89	3.91	4.27	4.12	3.83
SD	1.07	1.02	1.29	1.15	0.90	0.99	0.91
N	236	39	40	40	40	37	40
Romania							
Mean	4.35	4.37	4.19	4.18	4.34	4.51	4.60
SD	0.92	0.87	1.07	1.01	0.79	0.89	0.80
N	191	30	34	35	34	29	29
Diff.	0.27	-0.11	0.30	0.26	0.07	0.39	0.77
Prob.	0.005	ns	ns	ns	ns	ns	0.0005

The nurses in Bulgaria and the slum dwellers in Romania were the two groups that were demanding the largest risk mitigation. The lowest over all means for risk mitigation demand were obtained for the Bulgarian slum dwellers and the Romanian engineering students. A two way ANOVA-test, using the independent variables country and group showed that there were significant differences in means for the questions related to the demand for risk mitigation ($F = 2.25$, $p < 0.02$). The interaction was not significant ($F = 1.69$, ns). Figure 11 also shows correlation between the countries, but again the same three risk ratings differed: the risks to be assaulted, to be struck by lightning and to be attacked by wild dogs in the street.

Figure 11. **MEAN RATINGS OF DEMAND FOR RISK MITIGATION BY THE BULGARIAN AND ROMANIAN SAMPLES.**

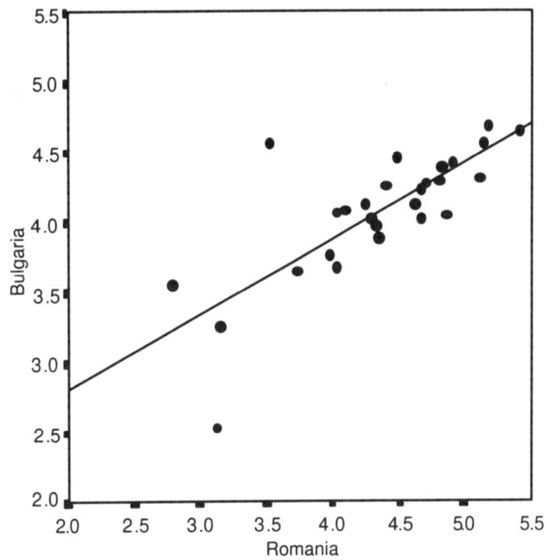

Mean ratings for protection possibilities in both countries are given in Table16. Differences in means between the two countries were not statistically significant, except for the group of workers ($p < 0.05$) and slum dwellers ($p < 0.001$), as it is shown in Table 16.

Table 16: **MEANS FOR THE QUESTIONS RELATED TO PROTECTION POSSIBILITIES**

Bulgaria							
	All respondents	Nurses	Workers	Students of engineering	Students of economics	Teachers	Slum dwellers
Mean	2.67	2.70	2.83	2.48	2.72	2.82	2.50
SD	1.04	1.14	1.16	0.83	0.73	1.21	1.11
N	237	40	39	40	40	38	40
Romania							
Mean	2.63	3.15	3.50	2.50	2.56	2.44	1.48
SD	1.17	1.15	1.47	0.67	0.79	0.81	0.87
N	191	30	34	35	34	29	29
Diff.	-0.04	0.45	0.68	0.02	-0.16	-0.38	-1.02
Prob.	ns	ns	0.035	ns	ns	ns	0.0005

A one way analysis of variance shows that the group differences in Romania were significant ($F = 14.94$, $p < 0.001$), but they were not significant among the groups in Bulgaria ($F = 0.84$, ns). The largest means for protection possibilities were obtained for the workers in both countries.

Those groups that were less secure about the protection possibilities against these risks were the Bulgarian students of engineering and the Romanian slum dwellers. A two way analysis of variance, using the independent variables country and group, showed that there were significant differences in means for the questions related to protection possibilities (F = 6.87, p < 0.001). The interaction between the two independent variables was significant (F = 6.03, p < 0.001).

The mean ratings of risk protection possibilities in Bulgaria and Romania were strongly correlated (see Figure 12). The figure shows also that the cross-national differences between means were not especially large.

Figure 12. **MEAN RATINGS OF PROTECTION POSSIBILITIES BY THE BULGARIAN AND ROMANIAN SAMPLES.**

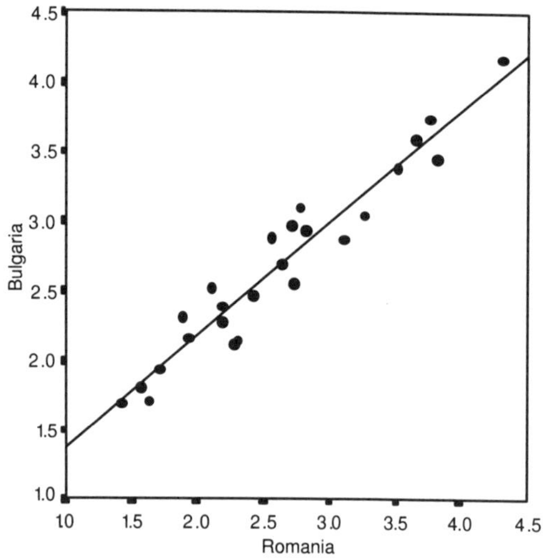

2.7 Probability of harm and severity of consequences

Table 17 below shows how the respondents in Romania and Bulgaria rated the personal probabilities of harm and the consequences of the hazards. Differences in means for all respondents between the two countries were statistically significant, as it is shown in Table 17. The difference between the means for slum dwellers and the means for workers was also significant. The other group differences were not significant. A one way analysis of variance shows that the group differences in Romania were significant (F = 5.78, p < 0.001), but they were not significant among the groups in Bulgaria (F = 1.39, ns).

The largest means for personal risk probability were obtained for the nurses in Bulgaria and for workers in Romania. The lowest probability ratings were found among the students of economics and engineering in both countries. A two way analysis of variance, using the independent variables country and group, showed that there were significant differences among all the studied groups, across the two countries (F = 5.40, p < 0.001). The two way interaction between the independent variables was significant (F = 4.08, p < 0.001). Means for general probability in both Bulgaria and Romania are given in Table 18. There were no statistically significant differences between the corresponding groups, in either country, except for slum dwellers, as it is shown in Table 18. A one way analysis of variance showed that there were statistically significant differences among the studied groups in Romania (F = 3.45, p < 0.01). A corresponding test was performed on the Bulgarian data, but the differences among the groups were not significant, (F = 1.57, ns). Nurses in Bulgaria and slum dwellers in Romania gave the highest ratings of general risk probability. The lowest ratings were given by nurses and workers in Bulgaria and by workers and students of engineering in Romania.

Table 17: **MEANS FOR THE PROBABILITY THAT THE RISK WILL CAUSE HARM TO THE RESPONDENT, PERSONALLY**

Bulgaria							
	All respon-dents	Nurses	Workers	Students of engineerins	Students of economics	Teachers	Slum dwellers
Mean	0.44	0.48	0.43	0.41	0.39	0.47	0.45
SD	0.14	0.16	0.13	0.10	0.12	0.17	0.14
N	128	19	15	30	23	23	18
Romania							
Mean	0.52	0.53	0.70	0.43	0.45	0.47	0.66
SD	0.22	0.17	0.28	0.13	0.17	0.28	0.15
N	107	16	15	22	25	15	14
Diff.	0.08	0.05	0.27	0.01	0.06	-0.01	0.21
Prob.	0.001	ns	0.003	ns	ns	ns	0.0005

Table 18: **MEANS FOR THE PROBABILITY THAT THE RISK WILL CAUSE HARM TO THE SOCIETY, GENERALLY**

Bulgaria							
	All respon-dents	Nurses	Workers	Students of engineering	Students of economics	Teachers	Slum dwellers
Mean	0.55	0.62	0.51	0.54	0.58	0.52	0.53
SD	0.15	0.11	0.15	0.12	0.12	0.20	0.14
N	112	21	25	10	17	17	22
Romania							
Mean	0.54	0.53	0.48	0.42	0.54	0.55	0.71
SD	0.21	0.16	0.21	0.15	0.12	0.20	0.26
N	84	14	20	12	10	13	15
Diff.	-0.01	-0.09	-0.03	-0.12	-0.04	0.02	0.17
Prob.	ns	ns	ns	ns	ns	ns	0.03

A two way variance analysis, using the independent variables country and group, showed that there were significant differences among all the studied groups, between the two countries ($F = 2.69$, $p < 0.004$). The two way interaction between the two independent variables was significant ($F = 3.10$; $p < 0.01$).

Personal severity ratings in both countries are given in Table 19. Differences in means for all respondents between the two countries were not statistically significant, except for the mean difference between the teachers, as it is shown in Table 19.

A one way analysis of variance shows that the group differences in Romania were significant ($F = 4.71, p < 0.001$), but they were not significant among the groups in Bulgaria ($F = 0.95$, ns). The largest means for personal risk consequences were obtained for the nurses in Bulgaria and for slum dwellers in Romania. The lowest consequence ratings were found among the students of economics and engineering in Romania and among workers and teachers in Bulgaria. A two way variance analysis, using the independent variables country and group, showed that there were significant differences among all the studied groups, across both countries ($F = 2.51, p < 0.005$). The two way interaction was not significant ($F = 0.92$, ns).

Table 19: **MEANS FOR SEVERITY OF CONSEQUENCES, PERSONALLY**

Bulgaria							
	All respon-dents	Nurses	Workers	Students of engineering	Students of economics	Teachers	Slum dwellers
Mean	4.13	4.41	4.23	4.00	3.97	4.35	3.90
SD	0.98	1.04	1.28	0.88	0.74	1.24	0.67
N	125	18	15	30	22	22	18
Romania							
Mean	4.36	4.56	4.84	3.93	4.04	4.94	4.27
SD	0.90	0.96	1.08	0.74	0.81	0.78	0.65
N	106	16	14	22	25	15	14
Diff.	0.23	0.15	0.61	-0.07	0.07	0.59	0.37
Prob.	ns	ns	ns	ns	ns	0.08	ns

Table 20: **MEANS FOR SEVERITY OF CONSEQUENCES, GENERALLY**

Bulgaria							
	All respon-dents	Nurses	Workers	Students of engineering	Students of economics	Teachers	Slum dwellers
Mean	4.11	4.50	3.72	4.03	4.48	4.02	4.05
SD	0.95	0.67	1.01	0.76	0.64	1.44	0.78
N	106	20	24	9	15	16	22
Romania							
Mean	4.41	4.63	4.14	4.29	3.90	4.66	4.84
SD	0.97	0.87	1.09	0.83	1.37	0.60	0.71
N	82	14	19	12	11	13	13
Diff.	0.30	0.13	0.42	0.26	-0.58	0.64	0.79
Prob.	0,41	ns	ns	ns	ns	ns	0.004

Figure 13. **MEAN RATINGS OF PROBABILITIES OF HARM, PERSONALLY, BY THE BULGARIAN AND ROMANIAN SAMPLES.**

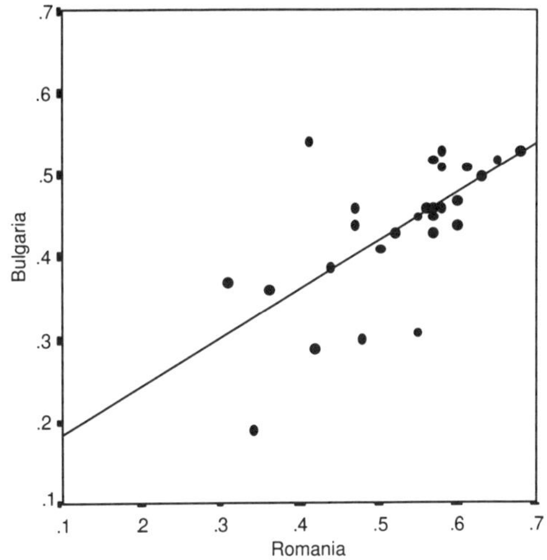

General severity ratings in Bulgaria and Romania are given in Table 20. There were statistically significant differences among all the respondents, in both countries, as it is shown in Table 20. A one way analysis of variance showed that there were statistically significant differences among the studied groups in Bulgaria (F = 2.00, ns). A corresponding test was performed on the Romanian data, but there were no significant differences among the groups (F = 1.53, ns). The largest consequence ratings were given by the nurses in Bulgaria and the slum dwellers in Romania. The lowest ratings for the severity of consequences were given by teachers and workers in Bulgaria and by students of economics and workers in Romania. A two way ANOVA-test, using the independent variables country and group, showed that there were significant differences in means for the questions related to risk consequences for people in general ($F = 2.20$, $p < 0.02$). The two way interaction was not significant (F = 1.86, ns). As it can be seen in Figure 13, the relationship between the means was rather weak.

There was also no strong linear relation between the Bulgarian and the Romanian means for the question of general risk probabilities (see Figure 14).

Figure 14. **MEAN RATINGS OF PROBABILITIES OF HARM IN GENERAL BY THE BULGARIAN AND ROMANIAN SAMPLES.**

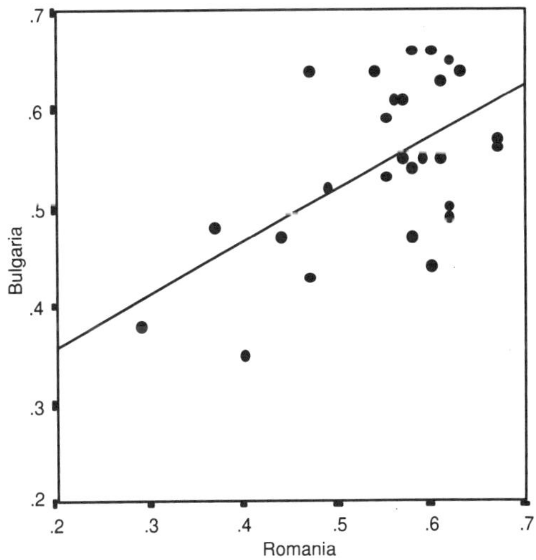

There seems to be a rather strong linear relationship between mean ratings for personal risk consequences when the Romanian and Bulgarian samples were compared (see Figure 15).

Further, there seems to be a rather strong linear relationship between mean ratings for general risk consequences when the Romanian and Bulgarian samples are compared (see Figure 16). However, there are three exceptions: to be attacked by wild dogs in the street, to become sick from food additives, and to be assaulted.

Figure 15. **MEAN RATINGS OF SEVERITY OF PERSONAL CONSEQUENCES BY THE BULGARIAN AND ROMANIAN SAMPLES.**

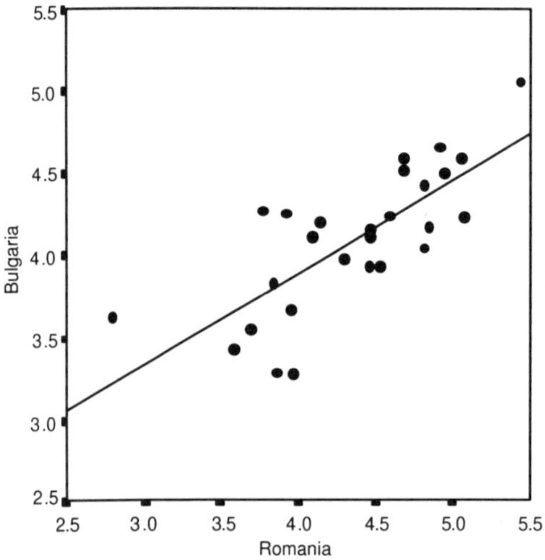

Figure 16. **MEAN RATINGS OF SEVERITY OF CONSEQUENCES IN GENERAL BY THE BULGARIAN AND ROMANIAN SAMPLES.**

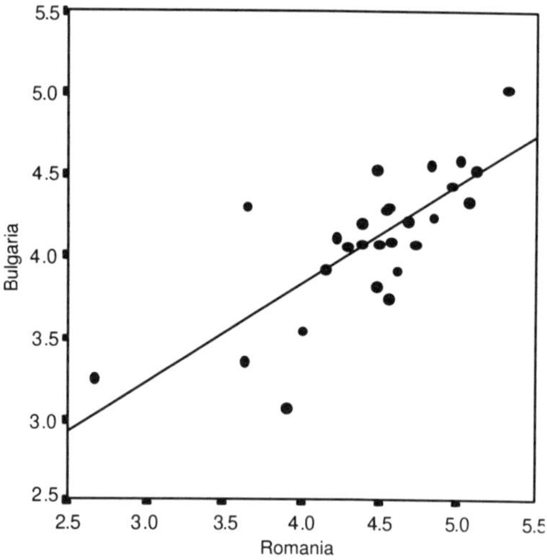

2.8 Influence on policy

Table 21 shows how the respondents in Romania and Bulgaria rated the possibilities to influence policy decisions. Differences in means between the groups in both countries were statistically significant, except for the teachers as it is shown in Table 21. A one way analysis of variance shows that the group differences in Romania were significant ($F = 15.83$, $p < 0.001$), but were not significant among the groups in Bulgaria ($F = 1.12$, ns). The workers in both countries felt that they had the largest possibilities to influence policy decisions. The two groups that felt the smallest possibility to exert influence were Bulgarian students of economics and Romanian slum dwellers. A two way ANOVA-test, using the independent variables country and group, showed that there were significant differences in means for the questions related to the possibilities to influence policy decisions ($F = 13.05$, $p < 0.001$). The two way interactions between the two independent variables were significant ($F = 11.40$, $p < 0.001$). Mean ratings are plotted in Figure 17.

Figure 17. **MEAN RATINGS FOR THE POSSIBILITIES OF INFLUENCING POLICY BY THE BULGARIAN AND ROMANIAN SAMPLES. THE SCALE ON THIS QUESTION IS INVERTED.**

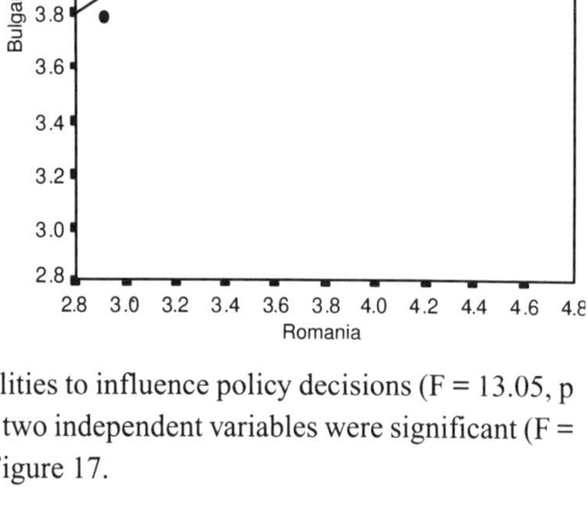

Table 21: **MEANS FOR THE QUESTIONS RELATED OF INFLUENCE POSSIBILITIES ON POLICY IN BULGARIA AND ROMANIA. A 6 ON THE SCALE MEANS NO POSSIBILITY TO INFLUENCE POLICY AT ALL**

Bulgaria							
	All respondents	Nurses	Workers	Students of engineering	Students of economics	Teachers	Slum dwellers
Mean	4.33	4.54	4.07	4.30	4.56	4.34	4.17
SD	1.15	1.17	1.54	1.08	0.74	1.32	0.87
N	234	38	40	40	38	38	40
Romania							
Mean	3.49	3.08	2.58	3.42	3.02	3.98	5.17
SD	1.50	1.38	1.74	1.03	1.29	1.20	0.53
N	187	30	34	33	34	28	28
Diff.	-0.84	-1.46	-1.49	-0.89	-1.55	-0.35	1.00
Prob.	0.0005	0.0005	0.0005	0.001	0.0005	ns	0.0005

2.9 Credibility of different risk information sources

Table 22 shows the results related to questions about credibility of risk information sources. The difference in means between the two countries was not statistically significant.

The highest levels of confidence in Bulgaria were found among nurses, followed by students of economics and engineering students. The lowest levels of confidence were found among the Bulgarian workers. A one way analysis of variance showed that there were also statistically

Table 22: **MEANS FOR THE CONFIDENCE IN DIFFERENT RISK INFORMATION SOURCES, IN BULGARIA AND ROMANIA**

Bulgaria							
	All respondents	Nurses	Workers	Students of engineering	Students of economics	Teachers	Slum dwellers
Mean	2.98	3.13	2.59	3.12	3.12	3.06	2.84
SD	0.70	0.55	0.72	0.65	0.67	0.82	0.62
N	239	40	40	39	40	40	40
Romania							
Mean	3.05	3.21	2.76	3.20	2.98	2.96	3.20
SD	0.66	0.56	0.56	0.64	0.56	0.71	0.83
N	190	30	34	35	34	28	29
Diff.	0.07	0.09	0.17	0.08	-0.15	-0.09	0.36
Prob.	ns	ns	ns	ns	ns	ns	ns

significant differences among the groups in Bulgaria ($F = 4.17$, $p < 0.001$). In Romania, the highest confidence levels were noted among nurses, followed by students of economics and teachers. The lowest were noted among Romanian workers. A similar ANOVA-test showed that the difference among groups in Romania was significant ($F = 2.48$, $p = 0.03$). A two way ANOVA-test, using the independent variables country and group, showed that there were significant differences in means for the questions related to credibility of risk information sources ($F = 2.87$, $p < 0.002$). The two way interaction was not significant ($F= 1.24$, ns).

The ratings of confidence in risk information sources in Bulgaria and Romania were similar, but there was one exception: friends and neighbors (see Figure 18). The Bulgarians seem to have had larger confidence in their friends and neighbors than the Romanian respondents.

Figure 18. **MEAN RATINGS OF CONFIDENCE IN RISK RELATED INFORMATION FROM DIFFERENT SOURCES BY THE BULGARIAN AND ROMANIAN SAMPLES.**

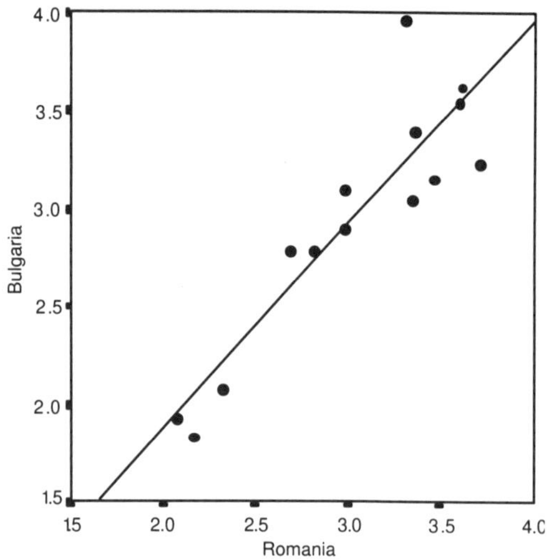

3 CONCLUSIONS

A first conclusion was related to the 100 risk questions derived from human activities, technologies and substances. The Bulgarian respondents assessed these risks as higher than the Swedish samples in Nyland's 1993 study, but they ranked the different risk questions in almost the same order (see Figure 1). In a comparison to Romania, the Bulgarians tended to value the risks higher than the Romanian respondents, but again the risks were ranked in almost the same order (see Figure 7).

When the Bulgarian data were compared to the Romanian ones for sections 2 to 5 in the questionnaires, the results showed some clear similarities as well as some differences in relation to risk perception, demand for risk mitigation, risk control, probability of harm, consequences of harm, possibilities to influence policy decisions, and finally in the credibility of risk information sources. For the questions related to the perceived risk levels, both personal and general, some differences that were noted in the data were connected to differences among groups in Bulgaria and Romania. These mean differences were significant between categories in Bulgaria but not in Romania. The means of the Romanian groups were quite close to each other. The difference in means between countries showed that personal risk perception in Bulgaria and Romania was quite similar (Figure 9). There were three exceptions: the risk to be assaulted, to be struck by lightning, and to be attacked by wild dogs in the street. It seems that the Bulgarian respondents were more concerned about the risk to be assaulted or attacked by wild dogs than the Romanians.

On the other hand, the Romanian samples were more concerned about the risk to be struck by lightning.

General risk levels were also perceived differently among the samples and between countries. These mean distinctions were significant, except for some of the mean differences between countries, and the rank order of general risks was almost the same for Bulgaria and Romania, except for the risks mentioned above (see Figure 11). The means of demand for risk mitigation and control over hazards did not diverge as much from one country to another. Most of the mean differences were not significant among the different samples in either countries, and the risks were all ranked in the same order (see Figure 12). The same results were obtained for the question related to protection possibilities where the correlation between the Bulgarian groups and the Romanian samples was rather strong (see Figure 12). The answers to the questions related to probability of harm and severity of consequences were much harder to analyze in the same way as the rest. The major reason for this difficulty may be the fact that the samples for these questions were rather small.

The answers to the questions about the possibilities to influence policy decisions showed significant differences between the countries, except for the answers provided by the teachers. Generally the respondents in Romania felt that they had larger possibilities to influence policy decisions than the Bulgarians. The same results were obtained for the credibility of the different risk information sources.

The results showed that the t-test of means for the different categories across countries tended to be insignificant for the different sections of the questionnaires. A more general conclusion related to both countries for all sections of the data material is that the six groups differed consistently in both countries in four cases, when it came to perceived risk and mitigation demand: nurses (high), workers (medium), economists (medium) and engineers (low). Teachers gave low ratings in Romania and high in Bulgaria. Slum dwellers gave high ratings in Romania and low in Bulgaria.

The results suggest, tentatively, that professional role and experience is important for risk perception. Several other studies in our unit support that conclusion, see introduction. The exception is the groups of teachers and slum dwellers and it would be interesting to study in-depth the reasons for the very different results in the two countries.

Gender was also an important variable in this study, especially in the comparison of the risk ratings in Bulgaria and Romania (see Figures 3, 4, 5, and 6). The results showed that men in Bulgaria were a diverging group that had a tendency to perceive risks as larger than the men in Romania did. One possible explanation could be that the Bulgarian men were older than the women, but that age difference, even though it was significant, was rather small (2.47 years). An ANOVA-test also showed that the independent variable age did not explain any of the mean differences in Bulgaria. The only significant variable in this case was the category to which the respondent was related.

The two countries‘ backgrounds were similar in several ways and mostly because they were both ex-communist countries in the Balkan area. A central distinction between them is that their cultural settings diverge.

A major difference was that the Bulgarians perceived the 100 first risks as larger than the Romanian and the Swedish samples did. Respondents in Romania, Bulgaria, Sweden and Brazil ranked the different risks in almost the same way, and the figures displaying the associations between those ratings display striking similarities between the studied countries. The above mentioned results could be discussed from the perspective of Cultural Theory.

A central idea in Cultural Theory is that people in their risk perceptions express cultural biases which in turn "support" different patterns of social relations. Several attempts have been made to investigate how large a part of risk perception could be explained by cultural aspects, but research on the topic shows diverging results.

According to Dake, his studies confirmed the general hypothesis of Cultural Theory. He claimed to have shown that societal concerns are predictable from cultural biases (Dake, 1991). On the other hand, Sjöberg‘s (1997, 1998) results do not support this hypothesis. Cultural Theory explained only a very minor share of risk perception, 5-10 percent. Contrary to expectations based on Cultural Theory, it was found here that the risk judgments made by the Bulgarian and Romanian samples followed similar rank orders even though the countries‘ cultures were different. This supports the notion that perceived risk is a function of real risk and perhaps media risk salience rather than culturally contingent values and beliefs (Sjöberg, 1996). French studies of risk perception also showed that the hypotheses of Cultural Theory should be rejected (Brenot & Bonnefours, 1995). The rank orders of perceived risk in this study are rather much the same, which suggests that the role of culture in risk perception is not as important as many authors have claimed earlier. Some of the differences could be explained by some kind of cultural bias or ecological real risk variation. One example is the important fear of lightning in Romania which was a risk that diverged from the rank orders for almost every section in part 2 of the questionnaire.

Earlier studies on the topic of cross-national comparison of risks have put forward a number of hypotheses to explain differences among countries. The first suggestion made was related to the effect of the country‘s size (Goszczynska et al, 1991). A Polish study showed a pattern of basic risk perception factors that were surprisingly similar to that of the American sample (Goszczynska et al, 1991), a pattern that remained unchanged even though tests were made to extend the traditional set of hazards. The mean risk ratings, on the other hand, showed clear differences among the studied countries. These differences in risk perception were explained by differences in size but could not be confirmed in the present study of risk perception in Bulgaria and Romania. Another hypothesis was related to the role of media. Research suggests that in countries with free reporting of risk items, risk perception will be higher than in countries with censored media. A study by Koné and Mullet suggests that media have a rather large impact on risk perception, and could therefore affect people‘s view of risks to an important extent (Koné & Mullet, 1994). But that paper gives only indirect evidence. A forthcoming paper of ours will explore media contents in Romania, Bulgaria, China and Sweden, see Sjöberg (in press) for some preliminary results.

It should be noted, however, that media effects on risk perception, if they exist, are probably not so strong and straightforward as often believed (af Wåhlberg & Sjöberg, in press). In the full report describing the present study (Sjöberg et al., 1996) we found media contents to be quite imperfectly related to risk perception. There were, e.g., hazards that were rated as quite high that media hardly treated at all, such as drinking tap water. In an extensive study of risk perception in Western Europe we found media contents to be somewhat, but rather weakly, related to the public´s risk perception (Sjöberg, Jansson & Viklund, 1999).

Some other questions that would have been of interest for this comparative analysis are connected to how Romanians and Bulgarians perceived risks during the communist period. It would have been quite interesting to have studied risk perception in 1985 in both countries and then compare those results to the results of the survey of 1993. Such a study would have helped in the task of identifying changes in risk perception due to the changes in e.g. media policy from 1985 to 1993. A remaining possibility we wish to explore in future work is that of archival "traces" of risk perception, i.e. various indirect remnants of behavior.

ACKNOWLEDGMENTS

The research reported here was supported by a grant from the Swedish Council for Research in the Humanistic and Social Sciences. The names of Kolarova and Rucai are given in alphabetical order; their responsibility is equal. Kolarova had main responsibility for the Bulgarian data; Rucai for the Romanian. Bernström worked with data analysis and wrote parts of the final report. Helena Flygelholm also took part in this work. For a more detailed report of the study, see Sjöberg et al. (1996).

4 REFERENCES

Abel, E. (Ed.). (1981). *The media in American society - What's news*. California: Institute of Contemporary Studies.

af Wåhlberg, A., & Sjöberg, L. (in press). Risk perception and the media. A review of research on media influence on public risk perception. *Journal of Risk Research*.

Barny, M.-H., & Brenot, J. (1991). La perception des situations à risque. *Preventiques. Prevention et gestion du risque, 41,* 11-19.

Bernström, M.-L. (1995). *Radiation, risk perception and risk communication*, (Rhizikon: Risk Research Report No. 25). Center for Risk Research: Stockholm School of Economics.

Blomqvist, A.-C., & Sjöberg, L. (1987). Risk and accident reports in the mass media. In L. Sjöberg (Ed.), *Risk and society* (pp. 207-218). London: Allen & Unwin.

Boholm, Å. (1998). Comparative studies of risk perception: A review of twenty years of research. *Journal of Risk Research*, 1, 135-164.

Brenot, J., & Bonnefours, S. (1995). *Approche socio-culturelle de la perception des risques* (Note SEGR/LSEES No. 95-17). IPSN.

Bryant, J., and Zillman, D. (Eds). (1994). *Media effects- Advances in theory and research.* New Jersey: LEA.

Combs, B., & Slovic, P. (1979). Newspaper coverage of causes of death. *Journalism Quarterly, 56,* 837-843, 849.

Cvetkovich, G., & Earle, T. C. (1994). The construction of justice: A case study of public participation in land management. In S. Opotow & S. Clayton (Eds.), Green justice, *Journal of Social Issues, 50*, 161-172.

Dake, K. (1990). *Technology on trial: Orienting dispositions toward environment and health hazards*. Ph. D. thesis, University of California, Berkeley.

Dake, K. (1991). Orienting dispositions in the perception of risk. *Journal of Cross-Cultural Psychology, 22*, 61-82.

Dominick, J. R. (1994). *The dynamics of mass communication*. New York: McGraw-Hill.

Douglas, M., & Wildavsky, A. (1982). *Risk and culture*. Berkely, CA: University of California Press.

Douglas, M. (1989). *Purity and danger - An analysis of the concepts of pollution and taboo.* Reading: Cox and Wyman.

Douglas, M. (1992). *Risk and blame. Essays in cultural theory*. London: Routledge.

Drottz-Sjöberg, B.-M. (1991). *Perception of risk- Studies of risk attitudes, perceptions and definitions.* Doctoral dissertation, Center for Risk Research, Stockholm of Economics.

Drottz-Sjöberg, B.-M., & Sjöberg, L. (1991). Attitudes and conceptions of adolescents with regard to nuclear power and radioactive wastes. *Journal of Applied Social Psychology, 21*, 2007-2035.

Dunlap, R. E., Gallup, G. H., Jr., & Gallup, A. M. (1993). Of global concern. Results of the health of the planet survey. *Environment, 35*, 7-39.

Dunwoody, S. (1994). Community structure and media risk coverage. *Risk- Issues in Health, and Safety*, *5*.

Eagly, A. H., & Chaiken, S. (1993). *The psychology of attitudes*. Orlando: Harcourt Brace Jovanovich.

Earle, T. C., & Cvetkovich, G. (1994). Risk communication: the societal construction of meaning and trust. In B. Brehmer & N. E. Sahlin (Eds.), *Future risks*. Amsterdam: Kluwer.

Engel, J., Blackwell, R., & Miniard, P. (1993). *Consumer behavior*, (7 th ed). Orlando: Dyrens Press.

Englander, T., Farago, K., Slovic, P., & Fischhoff, B. (1986). A comparative analysis of risk perception in Hungary and the United States. *Social Behavior, 1*, 55-66.

Flynn, J., Slovic, P., & Mertz, C. K. (1994). Gender, race, and perception of environmental health risks. *Risk Analysis, 14*, 1101-1108.

Freudenburg, W. R. & Pastor, S. K. (1992). Public responses to technological risks: Toward a sociological perspective. *The Sociological Quarterly, 33*, 389-412.

Friedman, S. M. (1994). The media, risk assessment and numbers, *Risk- Issues in Health and Safety,* 5.

Gerganov, E. N., Dilova, M. L., Paspalanova, E. P., & Petkova, K. G. (1995). Voting in the first post-totalitarian elections in Bulgaria. *Political Psychology, 16*, 737-748.

Goszczynska, M., Tyszka, T., & Slovic, P. (1991). Risk perception in Poland: A comparison with three other countries. *Journal of Behavioral Decision Making, 4,* 179-193.

Gooch, G. D. (1995). *Territories of environmental concern.* Linköping: Institute of Thema Research.

Gregory, R., Flynn, J., & Slovic, P. (1995). Technological stigma. *American Scientist, 83,* 220-223.

Izvorski, I. (1993). Economic reform in Bulgaria 1989-1993. *Communist Economic Transformation, 5*, 519-531.

Jeffres, L. W. (1994). *Mass media processes*, (2 nd ed). Illinois: Waveland Press.

Jianguang, Z. (1993). Environmental hazards in the Chinese public's eyes. *Risk Analysis, 13*, 509-514.

Kachelmeier, S. J., & Shehata, M. (1992). Examining risk preferences under high monetary incentives: Experimental evidence from the People's Republic of China. *American Economic Review, 82*, 1120-1141.

Karpowicz-Lazreg, C., & Mullet, E. (1993). Societal risks as seen by the French public. *Risk Analysis, 13*, 253-258.

Keon, C. F. (1989). Risk perception of Hong Kongese vs. Americans. *Risk Analysis, 9*, 401-405.

Kleihesselink, R. R., & Rosa, E. A. (1991). Cognitive representation of risk perceptions - A comparison of Japan and the United States. *Journal of Cross-Cultural Psychology, 222*, 11-28.

Koné, D., & Mullet, E. (1994). Societal risk perception and media coverage. *Risk Analysis, 14* (1), 21-24.

Kowalewski, D., & Porter, K. L. (1993). Environmental concern among local citizens - A test of competing perspectives. *Journal of Political and Military Sociology, 21,* 37-62.

Kraus, N., Malmfors, T., & Slovic, P. (1993). Intuitive toxicology: Experts and lay judgments of chemical risks. *Comment on toxicology, 4,* 441-484.

Leiss, W., & Chociolko, C. (1994). *Risk and responsibility*. Québec: McGill Queens.

Lichtenstein, S., Slovic, P., Fischhoff, B., Layman, M., & Combs, B. (1978). Judged frequency of lethal events. *Journal of Experimental Psychology: Human Learning and Memory, 4,* 551-578.

Lima, M. L. (1995). *Earthquakes are not seen in the same way by everyone. Cognitive adaptation and social identities in seismic risk perception.* Paper presented at the annual meeting of the Society for Risk Analysis (Europe), Risk Analysis and Management in a Global Economy, Stuttgart, Germany, 21-25 May 1995.

Lindstedt, U. (1995). Demokratin går långsamt. *Amnesty Press, 6*, 2-4.

Luthman, S. (1994). 1991-first quarter 1994 with initial revisions for 1993. *Key Economic Indicators, 2*(2), Stockholm Institute of East European Economics.

Marris, C., O'Riordan, T., & Simpson, A. (1995). *Redefining the cultural context of risk perception.* Paper presented at the annual meeting of the Society for Risk Analysis (Europe), Risk Analysis and Management in a Global Economy, Stuttgart, Germany, 21-25 May 1995.

Mechitov, A. I., & Rebrik, S. B. (1990). Studies of risk and safety perception in the USSR. In K. Borcherding, O. I. Larichev, & D. M. Messick (Eds.), *Contemporary Issues in Decision Making.*

Mihailescu, I. (1993). Mental stereotypes in post-totalitarian Romania. *Government and Opposition, 23* (3).

Natri, A. M. (1992). *Revolutionen som stals- demokratirörelsens syn på Rumänien efter Ceausescu.* Världspolitikens dagsfrågor, 9, Stockholm: UI.

Nelson, M. C. (1988). *Bitter bread- The famine in Norrbotten 1867-1868.* Stockholm: Almqvist & Wiksell International.

Nyland, L. G. (1993). *Risk perception in Brazil and Sweden* (Rhizikon: Risk Research Report No. 15). Center for Risk Research, Stockholm School of Economics.

OECD. (1992). *Bulgaria- An economic assessment.* Paris: OCDE.

OECD. (1993). *Romania- An economic assessment.* Paris: OCDE.

Peters, E., & Slovic, P. (1996). The role of affect and worldviews as orienting dispositions in the perception and acceptance of nuclear power. *Journal of Applied Social Psychology*, 26, 1427-1453.

Peters, H. P. (1994.) Mass media as an information channel and public arena. *Risk- Issues in Health and Safety*, *5* (3).

Pushev, O. (1991). The Bulgarian economy: transition or turmoil. In Ö. Sjöberg, & M. L. Wyzan (Eds.), *Economic changes in the Balkan States: Albania, Bulgaria, Romania and Yugoslavia.* London: Printer Publishers.

Renn, O. (1992). Risk communication: Towards a rational discourse with the public. *Journal of Hazardous Materials, 29*, 465-519.

Ronnås, P. (1991). The economic legacy of Ceausescu. In Ö. Sjöberg, & M. L. Wyzan (Eds.), *Economic changes in the Balkan States: Albania, Bulgaria, Romania and Yugoslavia.* London: Printer Publishers.

Rothman, S., & Lichter, S. R. (1987). Elite ideology and risk perception in nuclear energy policy. *American Political Science Review, 8*, 383-401.

Sjöberg, L. (1996). A discussion of the limitations of the psychometric and Cultural Theory approaches to risk perception. *Radiation Protection Dosimetry*, 68, 219-225.

Sjöberg, L. (1997). Explaining risk perception: An empirical and quantitative evaluation of cultural theory. *Risk Decision and Policy*, 2, 113-130.

Sjöberg, L. (1998). World views, political attitudes and risk perception. *Risk - Health, Safety and Environment*, 9, 137-152.

Sjöberg, L. (1999 a). Consequences of perceived risk: Demand for mitigation. *Journal of Risk Research*, 2, 129-149.

Sjöberg, L. (1999 b). Risk perception by the public and by experts: A dilemma in risk management. *Human Ecology Review*, 6 (2).

Sjöberg, L. (in press). Radiation risk perception in Western Europe. *Ambio*.

Sjöberg, L., & Drottz-Sjöberg, B.-M. (1991). Knowledge and perception of risk among nuclear power plant employees. *Risk Analysis, 11*, 607-618.

Sjöberg, L., & Drottz-Sjöberg, B.-M. (1994). *Risk perception of nuclear waste: experts and the public*, (Rhizikon: Risk Research Report No.16). Center for Risk Research, Stockholm School of Economics.

Sjöberg, L., Jansson, B., & Viklund, M. (1999). *Risk perception by the Swedish public 10 years after Chernobyl*. Unpublished manuscript.

Sjöberg, L., Kolarova, D., Rucai, A.-A., Bernström, M.-L., & Flygelholm, H. (1996). *Risk perception and media risk reports in Bulgaria and Romania* (RHIZIKON: Risk Research Report 23). Stockholm: Center for Risk Research.

Slovic, P., Fischhoff, B, & Lichtenstein, S. (1985). Characterizing perceived risk. In R. Kates, C. Hohenhemser, and J. Kasperson (Eds.), *Perillous progress: managing the hazard of technology*. Boulder: Westview Press.

Smith, A. H. (1991). The implications of change in East Central Europe for the Balkan socialist economies. In Ö. Sjöberg, & M. L. Wyzan (Eds.), *Economic changes in the Balkan States: Albania, Bulgaria, Romania and Yugoslavia*. London: Printer Publishers.

Snyder, L. B., & Rouse, R. A. (1995). The media can have more than an impersonal impact: The case of AIDS risk perceptions and behavior. *Health Communication, 7(2)*, 125-145.

Sowby, F. D. (1965). Radiation and other risks. *Health Physics, 11*, 879-887.

Starr, C. (1969). Social benefit versus technological risk. *Science*, *165*, 1232-1238.

Teigen, K. H., Brun, W., & Slovic, P. (1988). Societal risks as seen by a Norwegian public. *Journal of Behavioral Decision Making, 1*, 111-130.

Teodorescu, A. (1991). The future of a failure: the Romanian economy. In Ö. Sjöberg, & M. L. Wyzan (Eds.), *Economic changes in the Balkan States: Albania, Bulgaria, Romania and Yugoslavia*. London: Printer Publishers.

Thompson, M., Ellis, R., & Wildavsky, A. (1990). *Cultural theory, or why all that is permanent is bias*. Boulder, CO: Westview.

Waterstone, M. (1992). *Risk and society: the interaction of science, technology and public policy*, Kluwer academic publishers.

Wildavsky, A., & Dake, K. (1990). Theories of risk perception: Who fears what and why? *Daedalus, 119*(4), 41-60.

World Bank. (1991). *Bulgaria- Crisis and transition to a market economy, vol. 1*. Washington D.C.

World Bank. (1991). *Bulgaria- Crisis and transition to a market economy, vol. 2*. Washington D.C.

Wyzan, M. L. (1991). The Bulgarian economy in the immediate post-Zhivkov era. In Ö. Sjöberg, & M. L. Wyzan (Eds.), *Economic changes in the Balkan States: Albania, Bulgaria, Romania and Yugoslavia*. London: Printer Publishers.

5

THE COGNITIVE ARCHITECTURE OF RISK: PANCULTURAL UNITY OR CULTURAL SHAPING?

Eugene A. Rosa
Department of Sociology
Washington State University, Pullman, USA

Noriyuki Matsuda
Institute of Policy and Planning Sciences
University of Tsukuba, Tsukuba, Japan

Randall R. Kleinhesselink
Department of Psychology
Washington State University, Vancouver, USA

TABLE OF CONTENTS

SUMMARY

The intention of this study has been to test the hypothesis of "universalism of risk perception" using several student samples from the United States and Japan. The main objective was to reveal hidden patterns of risk perception based on psychometric characteristics among and between the different samples. In particular, the study was designed to demarcate between pancultural and culturally variant features of the cognitive processes of risks. One key element was the careful selection of a comparative culture. A comparison between Japan and a Western culture provides one of the most stringent tests of the hypotheses guiding the research program. Another key element was the development of a comparative instrument, in Phase I, which met a number of language translation criteria and tests of face validity.

Phase II of the research program produced a rich data set with which to test, not only the specified hypothesis, but also to perform sub-analyses and other refinements. This data set was first analyzed using the standard factor-analytic techniques of the standard psychometric paradigm. In general, the findings from the United States-Japan comparison replicated previous results produced within this paradigm - including the cross-cultural results. The cognitive maps produced by this analysis indicated common contours between the Americans and Japanese, but different cognitive content. Without a formal basis for judging whether contour or content should be the criterion for determining which of our hypotheses were supported, the fundamental questions generating the research program remained unresolved.

To resolve them we turned to latent variable modeling in Phase III of the research program, still in progress. Thus far this phase has produced preliminary results on only the sub-set of seven nuclear risks contained in the data set. These results are, nevertheless, promising. We developed a refined, best-fitting model for each culture based upon the nuclear items. What these models seem to be telling us is that Americans and Japanese have similar images of dread and catastrophic potential underlying their perceptions of nuclear risks. But, importantly, the cognitive routes they use to arrive at those images are fundamentally different.

1 INTRODUCTION

There are a variety of time worn, compelling reasons for a cross-cultural approach to the serious study of perceptions - including, of course, risk perceptions. It is, first, this context that provides the "natural laboratory" for pursuing the long-standing intrinsic interest in human variation - in all of its many forms. Indeed, a mapping of the full range of human variation has occupied the attention of cultural anthropology for over a century. Second, cross-cultural comparisons provide a context for extending and evaluating research protocols. Generalizations about some aspect of behavior may be restricted because they are bound to a particular methodology - the so-called method effect. Third, some topics, such as risk, are so closely tied to policy concerns that there is a need for their greater understanding in order to facilitate international dialogue and agreement.

While these reasons - and others left unmentioned - have been sufficiently compelling to stimulate generous volumes of cross-cultural research, there is, in our view, an even more compelling reason to understand risk cross-culturally. A context of cultural variability provides the ideal setting for uncovering fundamental human processes - perceptual, analytic, emotional, behavioral - that may overpower the deep and pervasive influence of culture. As a single species it is reasonable to expect human universals for certain fundamental cognitive processes. An understanding of these processes can provide the requisite foundation for developing more generalizable models of human thought and behavior by demarcating between species based and cultural based influences on human thought and action.

A cross-cultural orientation is especially applicable to an understanding of risk perceptions. The idea of risk taps into one of the most universal of human experiences: uncertainty. Humans have always and will always live in a world comprising of both knowledge and uncertainty. Indeed, with the work of Heisenberg (1958) and others, the weight of theoretical opinion now concludes that uncertainty is inescapably embedded in our most sophisticated knowledge system-science. Indeed, uncertainty is likely embedded in all knowledge systems.

Risk, likewise, embeds uncertainty. Nearly all definitions of risk treat uncertainty as the *sin qua non* of this broad concept. One of the most explicit recognition of the role of uncertainty in the idea of risk can be found in the definition by Rosa (1998):

> *Risk is a situation or event where something of human value*
> *(including humans themselves) has been put at stake and where*
> *the outcome is uncertain.*

To summarize, uncertainty is a condition of universal ubiquity - across time and space - on the one hand, and an indispensably defining feature of risk, on the other hand. Thus, risk is also a universal human condition - across time and space. This ontological foundation leads to a pair of competing hypothesis to guide research:

(i) Risks are expected to be perceived in the same basic way across cultures because humans - as a common species - have a common cognitive architecture that filters perceptions in the same way; or

(ii) Human perceptions (including perceptions of risk or uncertainty) are so embedded within cultural meanings that they are expected to exhibit the same variability as for other aspects of human experience?

Hypothesis (i) derives from the recognition of the unique qualities of the human species, especially qualities that reveal the complex cognitive structures common to humans everywhere. As a single species it is assumed that humans (homo sapiens) have a common cognitive architecture and are capable of processing complex inputs along multiple channels. Given a common cognitive architecture, we would expect that a common set of neurological algorithms is used by humans everywhere to process perceptions, thoughts, memory and to make judgments and decisions. Furthermore, as aptly expressed by Nyland, "These aspects of our experiences will be stored in our memories, in a kind of structure, or a cognitive map of reality. A cognitive map is understood as a kind of structure, that is an abstract representation of events, objects, and relationships in the real world" (1993:2).

Hence, with proper stimuli and with proper techniques it should be possible to tap into the cognitive architecture and represent its stored structures as cognitive maps. Cognitive psychology has developed such stimuli and techniques for the mapping of perceptions, including risk perceptions. Because of the common cognitive architecture underlying all human thought, the research hypothesis predicts cognitive maps that are culturally invariant.

With *hypothesis (ii),* cultural theory challenges the expectation that risk perceptions are invariant across cultures. While it may be undeniably true that the human species is partly defined by a common cognitive architecture, it doesn't necessarily follow that that architecture will process the multitude of environmental inputs in a singular way. Instead, that architecture -a raw, unprogrammed structure- may be little more than a neurological readiness for complex reasoning. Cultural theorists such as Douglas and Wildavsky (1982) and Thompson, Ellis, and Wildavsky (1990) would argue, requires programmed meaning - the very meaning that is provided by culture.

An analogy between computer hardware and software may be useful here. The cognitive architecture of the humans species may be compared to that of a computer (i.e., hardware); it has a "readiness" to perform complex tasks, but is totally incapable of performing those tasks without instructions - a program, or software. "Software," in the world of ordinary human experience, is culture. Culture imposes an organized structure on the individual's, highly capable, but unprogrammed psyche. It imposes a template (a cultural map), shared by other members of the culture, that filters environmental inputs in systematic ways. Culture, therefore, filters perceptions - including risk perceptions. And since cultures vary widely, cultural theory predicts that risk perceptions will exhibit a parallel variability. Furthermore, not only will perceptions vary by culture, but they will also vary by group identification for those individuals living in complex, modern cultures (Douglas and Wildavsky, 1982).

The stated hypotheses point directly to cross-cultural methodology as the most apt means for addressing them. The remainder of this chapter discusses a cross-cultural research program based at Washington State University (WSU) devoted to demarcating between pan-cultural elements and culturally variable elements of risk perceptions. The substantive focus of the research program is perceptual comparisons between Americans and Japanese, and the program comprises three distinct phases: Phase I, a pre-test phase, Phase II, a execution stage, and Phase III, a modeling stage. Before describing the research program in detail we review other cross-cultural studies of risk perceptions. It is to that task that we now turn.

2 CROSS-CULTURAL STUDIES

With few exceptions cross-cultural studies of risk perceptions have adopted the psychometric paradigm developed by Paul Slovic and his colleagues at Decision Research in Eugene, Oregon USA (Lichtenstein, Slovic, Layman & Combs, 1978; Slovic, Fischhoff & Lichtenstein, 1985; Slovic, 1987). The psychometric paradigm emphasizes the use of psychophysical scaling and multivariate techniques (especially factor analysis) to create quantified cognitive risk maps. The cumulative empirical findings have led to two key generalizations. First, laypersons' perceptions of risk are systematically biased; they typically exaggerate risks associated with infrequent, catastrophic, and involuntary events, and underestimate risks associated with frequent, familiar, voluntary events. Second, laypersons - especially in comparison to risk experts - emphasize a variety of qualitative rather than technical or quantitative features of the common risks they face. Yet, despite the complexity introduced by the qualitative features, and despite the rapidly growing number of risks salient to laypersons, risk perceptions can be mapped via factor analysis with a relatively simple structure. In particular, a two factor structure typically accounts for a clear majority (roughly 70% to nearly 90%) of the variance in risk perceptions. The first factor is labeled *dread* risk and the second is labeled *unknown* risk.

Using psychometric techniques comparable to those applied to U.S. samples, risk perception data have, to date, been collected in Hungary (Englander, Farago, Slovic, & Fischhoff, 1986), Norway (Teigen, Brun, & Slovic, 1988), Hong Kong (Keown, 1989), Poland (Goszczynzka, Tyszka, and Slovic, 1991), France (Karpowicz-Lazreg and Mullet 1993), and Sweden and Brazil (Nyland, 1993) . Most striking is the general finding that for each of the sampled cultures, just as for the U.S., a two factor structure emerges. Moreover, almost without exception the two factors are most aptly labeled dread and unknown, again displaying consistency with the U.S. findings. In very general terms, then, people in diverse cultures apparently characterize their risk environments in similar ways.

As favorable as this evidence may be for claims of universal cognitive processing of risk stimuli there are good reasons to exercise caution before reaching such a conclusion. First, despite the apparent similarity in the overall cognitive maps, there were a number of differences in the perception of individual risks. For example, while Americans in these studies were often most concerned with unknown risks, members of the other cultures were often most concerned with known risks. Second, virtually all cross-cultural studies cited are weakened by a common methodological flaw[1] : the failure to collect U.S. data concurrently with the collection of the comparative cultural data. Instead, more recent data from the other cultures are compared to U.S. data that is typically over a decade old. This practice is tenuous because it assumes that the risk perceptions of Americans are stable and, therefore, appropriate for comparative analysis. Available evidence, however, suggests that American risk perceptions have been changing (Kraus and Slovic, 1988; Keown, 1989). Third, all comparisons are with societies that are either relatively small (geographically and in population size), such as Hong Kong and Hungary, or are part of a Western cultural tradition. These methodological practices raise further questions about the true comparability of the samples examined.

JAPAN: PARADIGM COMPARATIVE CULTURE

To avoid the limitations of previous cross-cultural studies of risk perception requires (i) the collection of data in two cultures simultaneously, and (ii) the careful selection of the cultures to

[1] A clear exception if the study comparing risk perceptions of citizens in Sweden and Brazil (Nyland, 1993).

[2] See Brislin (1993) for a corroboration and elaboration of this argument.

be compared. The later requirement comprises two key conditions: first, the two cultures should be vastly different, not part of the same cultural tradition, and, second, they must be exposed to or share many risks in common. These two conditions, we argue, can be met with a comparison between Japan and virtually any large, advanced Western society.

That Japanese history and culture is uniquely different from Western cultures is a fact that virtually no one would dispute. Second, it has a large population (at 126 million, the seventh largest in the world), and, like many societies in the West., it has clearly reached a stage of advanced industrialization with the consequence that its citizens are exposed to many of the same risks as citizens in the West. A comparison between cultures with such distinct cultural traditions, between West and East, is a more stringent test of the universality hypothesis than, for example, a comparison between two Western cultures. For these, and other reasons, the WSU Cross-Cultural Research Program has focused on comparisons between Americans and Japanese.

3 RESEARCH PROGRAM

3.1 Research design

The WSU Cross-Cultural Research Program comprises three distinct phases: Phase I, a pre-test phase, Phase II, an execution stage, and Phase III, a latent variable modeling stage.
Table 1 summarizes the research program by phase.

SAMPLES

College student samples were used in all phases of the research program. It is appropriate to note that college student samples, often viewed as a methodological shortcoming in research, can be viewed here as a special strength for several reasons. First, because the central concept under investigation is culture, by definition this means that any group sampled within the culture should bear its influence. Second, many of the risks being assessed are the result of recent developments in science or high technology. Proper assessment of these risks, then, requires some awareness on the part of respondents, and this is more assured with college students who are generally more informed on such matters. Third, college students are expected to have stronger cognitive strategies focusing on perception, memory and thinking (Sears, 1986). Fourth, and most importantly, sociologists (Form, 1979) have identified a major concomitant of advanced industrialization (and the internationalization of economies) to be a "convergence" in the overall structure of societies, accompanied by growing similarities in attitudes and values. Colleges and universities are likely one setting where the impacts of changing values and attitudes are first experienced. This means, in turn, we would expect college student samples from two advanced societies, *ceteris paribus*, to be more similar than samples drawn from other settings. The net effect of this expectation is a more conservative, and therefore more rigorous, test of cultural differences[2].

Table 1. **WASHINGTON STATE UNIVERSITY (WSU) U.S. - JAPAN CROSS-CULTURAL RISK PERCEPTION RESEARCH PROGRAM**

	Principal Objectives:	Samples	Sample Sizes	Product
PHASE I (Pre-Test)	1. Develop Japanese Language Version of Psychometric	U.S: Students at Washington State University	62	Validated Instrument and Viable U.S. - Japan Research Protocols
	2. Develop Research Protocols	Japan: Students on a Short Stay in the U.S. for Veterinary or Intensive Language Training	69	
	3. Pre-Test Instruments and Protocols			
PHASE II (Execution)	1. Verify Japanese Translation of Instrument	U.S: Students at Washington State University	152	Completed Data Set From Concurrent Administration in the U.S. and in Japan
	2. Administer Completed Instrument Concurrently to Samples in the U.S. and Japan	Japan: Students at Nihon University and University of Tsukuba	273	
PHASE III (Modeling)	1. Subject Data Collected Collected in Phase II to Latent Variable Models	Same Samples as for Phase II	Same as Phase II	Models Development in Progress
	2. Make a more Precise Demarcation between Cultural Invariant and Variant Elements of Risk Perception			

3.2 Phase I: Pre-test:

The principal objectives of this phase were to create a Japanese language version of the standard psychometric instrument used in previous studies, and to refine the research protocols for administering the instrument.

INSTRUMENT DEVELOPMENT

Instrument development took place in the following steps. We began with the taxonomy of 81 risks and hazards employing 9 dimensions of risk evaluation that had been standardized and tested rigorously by Slovic et al. (1985). We believed that a refined version of this instrument would permit valid, direct comparisons of risk perceptions between Americans and Japanese and to provide a baseline for future comparisons. Slovic et al. (1985) later expanded the dimensions to 18, but the added dimensions are far less empirically established. Next, the standard list of 81 risks and nine dimensions were modified to this cross-cultural context. Risks that did not exist in both cultures (e.g. high school football) or did not have comparable rates of incidence or familiarity to both cultures (e.g. coal tar hair dyes) were removed. To preserve the factor space

Table 2. **LIST OF 70 RISKS AND HAZARDS FOR CROSS-CULTURAL RESEARCH**

Consumer Products	Environmental Risks
1. Microwave Ovens	1. Fluoridated Water
2. Food Preservatives	2. Lead Paint
3. Saccharin	3. Leaded Gasoline
4. Caffeine	4. Electric Fields
5. Chemical Cleaners	5. Nitrogen Fertilizers
6. Electrical Appliances	6. 2,4,5-T (Dioxins)
7. Food Irradiation	7. Pesticides
8. Home Space Heaters	8. Asbestos
	9. Mercury
Social and Medical Risks	10. Greenhouse Effect
1. Prescription Drugs	11. Auto Exhaust Fumes
2. Oral Contraceptives	12. LNG Transport
3. X-rays	13. Herbicides
4. IUDs	14. Upper Ozone Depletion
5. Antibiotics	
6. Aspirin	**Non-nuclear Technologies**
7. Vaccinations	1. Hydroelectric Power
8. DNA Research	2. Plastics (manufacture/disposal)
9. Crime	3. Synthetic Rubber Manufacture
10. Surgery	4. Solar Electric
11. Alcoholic Diseases	5. Fossil Fuel Electric Power
12. Radiation Therapy	6. Heavy Farm Machinery
13. Alcohol Related Accidents	7. Elevators
14. Motor Vehicle Accidents	8. Bridges (construction/use)
15. AIDS	9. Coal Mining Diseases
16. Handguns	10. Dams (construction/use)
	11. Skyscraper Fires
Voluntary Risks	12. Coal Mining Accidents
1. Skateboards	13. Non-nuclear Electric Power
2. Smoking Tobacco	14. Skyscrapers (build/maintain)
3. Trampolines	15. Dynamite
4. Downhill Skiing	
5. Bicycles	**Nuclear Technologies**
6. Motorcycles	1. Transport Nuclear Materials
7. Fireworks	2. Nuclear Power Routine Radiation
8. Hang Gliding	3. Radioactive Waste Disposal
9. Railroad Accidents	4. Nuclear Reactor Accidents
10. Commercial Aviation	5. Nuclear Weapons Testing
	6. Nuclear Power Production
	7. Nuclear Weapons (war)

characteristics of the Slovic et al. cognitive maps, approximately 6 risks were removed from each of the 4 quadrants of that factor space. In addition to the 3 nuclear issues included in the typical Slovic et al. instrument, 4 additional nuclear risks (radioactive waste disposal, transportation of nuclear waste, nuclear power plant operation, and food irradiation) were included. Finally, newer cross-cultural risks (upper atmosphere ozone depletion, the greenhouse effect, dioxins, crime, pesticides, herbicides and AIDS) never before examined were added to the list. These modifications produced a final list of 70 risks. Table 2 is a complete list of the 70 risks.

ADMINISTRATION PROTOCOL

The completion of the risk questionnaire is a somewhat tedious and demanding task, susceptible to effects of fatigue and inattention. To reduce this susceptibility we sought to keep the amount of time necessary to complete the survey under 60 minutes. To accomplish this, only seven of the nine standard dimensions of risk evaluation were retained for this survey. Four of the risk dimensions retained (controllable-uncontrollable, calm-dread, not catastrophic-catastrophic, and voluntary-involuntary) were associated with the Dread-risk Factor in the typical Slovic et al. study, and the other 3 risk dimensions (known-unknown to the individual, old-new, and known-unknown to science) were associated with the Unknown-risk Factor. These 7 dimensions of risk perception have been identified by Slovic (1987, 1988) as having high factor loadings in previous research. The risk dimensions were described in a short question format and 7-point Likert-type scales were anchored by the descriptive bi-polar adjectives (e.g., dread-calm, new-old). The dimensions and response formats are presented in Table 3.

Table 3. **DIMENSIONS OF RISK PERCEPTION AND RESPONSE FORMAT**

1. **New/Old**
   Is this risk new and novel or old and familiar?
   New ... Old ... Omit
   1 ------- 2 ------- 3 -------- 4 ------- 5 -------- 6 ------- 7 ( )

2. **Scientific Knowledge**
   To what extent are the risks known to science?
   Unknown ... Known ... Omit
   1 ------- 2 ------- 3 -------- 4 ------- 5 -------- 6 ------- 7 ( )

3. **Individual Knowledge**
   To what extent are the risks known precisely by the persons exposed to those risks?
   Unknown ... Known ... Omit
   1 ------- 2 ------- 3 -------- 4 ------- 5 -------- 6 ------- 7 ( )

4. **Involuntary/voluntary**
   Do people become exposed to the risk voluntarily?
   Voluntary ... Involuntary ... Omit
   1 ------- 2 ------- 3 -------- 4 ------- 5 -------- 6 ------- 7 ( )

5. **Controllability**
   If you are exposed to the risk, to what extent can you, by personal skill or diligence, avoid death?
   Not Controllable ... Controllable ... Omit
   1 ------- 2 ------- 3 -------- 4 ------- 5 -------- 6 ------- 7 ( )

6. **Dread**
   Is this a risk that people have learned to live with and can think about calmly, or is it one that people have a great dread for - on the level of a gut reaction?
   Thoughts of Dread ... Thoughts of Calm ... Omit
   1 ------- 2 ------- 3 -------- 4 ------- 5 -------- 6 ------- 7 ( )

7. **Catastrophic**
   Is this a risk that may harm people one at a time (noncatastrophic) or a risk that harms large numbers of people at once (catastrophic)?
   Catastrophic ... Non-catastrophic ... Omit
   1 ------- 2 ------- 3 -------- 4 ------- 5 -------- 6 ------- 7 ( )

JAPANESE TRANSLATIONS

After constructing the English language version of the questionnaire, two Japanese native language translators were employed to translate the survey into standard Japanese language usage. Working independently, one translator provided a Japanese translation. The second translator back-translated this version into English. This procedure was followed until the back-translated version provided the same common meaning as the initial English version. As a further check two additional native Japanese translators revised and confirmed the accuracy of the translations. The translations were subject to further review and corroboration in Tokyo by native Japanese speakers at the Institute of Applied Energy and by risk experts at the National Institute of Police Science. A final confirmation of the accuracy of the translations and of the usefulness of the instrument was its adoption by the Institute of Applied Energy in Tokyo in surveys it conducted.

PRE-TEST

Two samples of college students were chosen for the initial evaluation of the revised psychometric instrument. See Table 1. The first sample of 62 (37 female and 25 male) U.S. college students, enrolled in upper-division social science classes at Washington State University, (Pullman and Vancouver campuses), completed the survey during the Summer term. The comparable sample of 69 (21 female and 47 male) Japanese college Students also completed the survey during the same period. This second sample consisted of two separate groups, approximately equal in size, of Japanese students from a variety of Japanese universities temporarily located in the U.S. One group was enrolled in intensive English language instruction at Washington State University. They had recently arrived from Japan, had no previous exposure to American culture, and had low mastery of the English language. The other group consisted of Japanese veterinary students, who, as part of their training, attended clinical workshops in the U.S.: one week at University of California-Davis and one week at Washington State University. Upon completion of the survey virtually every participant, with the aid of native speaking Japanese translators, was debriefed about their comprehension of the survey instrument. Of the 69 Japanese participants only four raised any questions about the meaning of survey items, and even here the discrepancies were too slight to seriously vitiate their substantive interpretations. All Japanese participants received $10 in remuneration for completing the survey, while U. S. participants received additional points toward their course grade.

This pre-test confirmed our two key expectations. We were able to administer the instrument well within the hour time limit target we had imposed: the vast majority of students from both cultures completed the instrument well within the hour, and none went over the hour. Based upon the feedback from the primary translators, from the native language reviewers, and from the debriefing of subjects after administering the instrument, we were confident that we created a common instrument (in two languages) that would produce valid results in both cultures.

RESULTS

We performed a variety of analyses on this pre-test data. Since the results of those analyses have already appeared elsewhere (Kleinhesselink and Rosa, 1991) we will not re-produce them in detail here. Instead, we summarize the key findings.

Frequency distributions, including means and standard deviations, were computed for each of the 70 risks and for each of the seven dimensions of risk. The Japanese sample tended to view the 70 risks and hazards as *older* (33 vs. 16 significant t-tests) and risks for which they had *more individual knowledge* (32 vs. 12 significant t-tests). In contrast, the U.S. sample tended to view the 70 risks as *more controllable* (21 vs. 13 significant t-tests) *less dreaded* (22 vs. 7 significant t-tests) and *less catastrophic* (22 vs. 9 significant t-tests).

Following the procedures of Slovic et al. (1985) a correlation matrix of the responses on the seven dimensions of risk was obtained. These dimensions were factor analyzed with a principal components method. For both samples a two factor solution emerged, similar and consistent with all previous studies using the procedures of the psychometric paradigm. And for both samples, the new-old, individual knowledge, and scientific knowledge dimensions had factor loadings that ranged from .73 to .84 while none of the other 4 dimensions had a loading higher than .44. This factor, previously identified as the Known-Unknown risk factor, explained 26% of the averaged variance. Also for both samples, the catastrophic potential, controllability, voluntariness and dread dimensions had factor loadings ranging from .52 to .76 while none of the other 3 dimensions had loadings higher than .26. This factor has previously been identified as the Dread risk factor, and explained 33% of the averaged variance.

Taken together, the Dread factor and the Unknown factor, explained 60% of the variance in the American sample and 56% of the variance in the Japanese sample. While the amount of explained variance is somewhat lower than in previous studies, it is substantively consistent with those studies.

Following the procedures of Slovic et al. (1985) two factor scores, a *dread* score and an *unknown*, score were computed for each of the 70 risks by weighting the ratings on each risk dimension proportionately to the importance of the dimension for the factor and summing over all dimensions. This weighted summation assigned individual risks items high or low scores dependent upon whether the ratings are high or low on the dimensions most closely associated with each of the two factors. The results were plotted in a two-dimension factor space, producing a cognitive map for each sampled culture. The overall contours of the cognitive maps were quite similar for both cultures, suggesting that the "cognitive map" of risk perceptions of environmental hazards has the same coordinates for both cultures. Furthermore, the American and Japanese contours were quite similar to the cognitive maps produced in previous studies.

While there were striking similarities between the two cultures - a common two factor solution, and similar cognitive maps - there were also striking differences. The two cultures were most similar in the way in which dread was assigned to their perceptions of environmental hazards. For example, the most dreaded activity in both cultures was nuclear war. Additionally, in both cultures all seven nuclear issues (nuclear power production, nuclear weapons (war), transportation of radioactive materials, nuclear waste disposal, reactor accidents, nuclear weapons tests, and routine emissions of radioactive gases) were closely clustered together and highly dreaded.

However, while the clustering of nuclear issues at the extreme end of the dread dimension was similar between the two cultures, the clustering on the Known-Unknown risk dimension was quite different. To further explore these differences we computed the mean nuclear risk scores for both cultural samples on the three dimensions associated with the unknown risk factor. In the U.S. sample, nuclear risks were viewed as generally unknown both to the individual and to science, and somewhat new. In contrast, these same nuclear risks were viewed as some of the most well known risks to the Japanese sample; older and better known to the individual or to science. For example, the average individuals knowledge score of the 7 nuclear risks was 3.07 for the Japanese sample and 5.84 for the U.S. sample on the 1-7 Likert scale. Likewise, the scientific knowledge average score was 3.84 for the Japanese sample and 6.31 for the U.S. sample. (For the knowledge dimension a high score means "unknown.") The old-new average score for the Japanese sample was 6.24 vs. 3.63 for the U.S. sample. (For the old-new dimension a high score means "old.") All 21 of the t-test comparisons for the three dimensions were significant. Furthermore, on a 7 point Likert scale, a score of 4 is neutral. Thus, in both a relative comparison between the two cultures and in the Likert scale meaning comparison, Japanese citizens viewed nuclear issues as somewhat known while U.S. citizens viewed them as very much unknown.

Similarly, the Japanese students rated many of the medical and chemical risks as significantly older and more known than did the American Students. At the same time, the Japanese students perceived a variety of voluntary risks (bicycles, downhill skiing, hang gliding, fireworks, skateboards, trampolines, electric appliances, etc.) as less known than did their American counterparts.

DISCUSSION

The principal objectives of Phase I of the WSU Research Program - to create a Japanese language version of the standard psychometric instrument used in previous studies, and to refine the research protocols for administering the instrument - were effectively achieved. The final, refined instrument passed several levels of scrutiny regarding the validity of translation, including that of the Japanese subjects who, during the debriefing, raised few issues about the meaning of the instrument's instructions or its items. The protocols for administering the instrument were, likewise, refined and proved to be effective and practical. The pre-test results, especially for the American sample which could be compared directly to previous studies, also attested to the effectiveness of the refined instrument. The fact that the results were comparable to prior findings suggests that the instrument was tapping the same perceptual structure as the original instrument from which it was derived.

3.3 Phase II: Execution of data collection

The first objective of this stage was to collect samples of sufficient size, for a variety of multivariate analyses, in both cultures simultaneously. The second objective, was to subject the data to the analyses typical of the psychometric tradition. The key output of these analyses would be cognitive maps for the American and Japanese samples that could be compared to (i) the pre-test results for consistency, and (ii) to prior American and other cross-cultural findings.

DATA COLLECTION

The risk instrument developed in Phase I was used to collect all subsequent data collected in the WSU Research Program. It was administered concurrently to samples of American and Japanese university students. A total of 152 U.S. male and female university students enrolled in psychology or sociology courses at Washington State University completed the risk instrument during a Fall semester. During the same semester, 273 Japanese male and female university students enrolled in psychology or social science courses at either Nihon University of the University of Tsukuba also completed the risk instrument. Concurrently collected data, absent from nearly all previous cross-cultural studies, was an important element of this data collection phase. It is important because mishaps, accidents, and disasters - and the media they attract - could contaminate differences found between cultures measured at two different points in time. Found differences might not be due to the shaping of culture, but, instead, might be an artifact of historical events.

RESULTS

The data were first subjected to a straightforward comparison. The population means and standard deviations were computed for each of the 70 risks and for each of the seven dimensions of risk. Table 4 presents these means and standard deviations for the two samples on each of the seven dimensions after averaging respondent responses on all 70 risks and hazards.

Table 4 also presents the number of significant t-tests ($P<.05$) between the American and Japanese samples for each of the 70 risks and hazards evaluated within each of the seven dimensions of risk perception.

Patterns in this data resemble those found in the analyses of the pre-test data of Phase I. In general, the Japanese sample tended to view the 70 risks as *older* (33 vs. 14 significant t-tests) and as risks for which there were *less scientific knowledge* (35 vs. 21 significant t-tests). There was no overall difference in individual knowledge (29 vs. 20 significant t-tests). A closer examination of the seven nuclear technology-related hazards (transporting nuclear materials, nuclear power-routine emissions, radioactive waste disposal, nuclear reactor accidents, nuclear weapons testing, nuclear power production, and nuclear weapons [war]), revealed a striking pattern of cultural differences; namely, that nuclear risks were perceived as *well-known* risks to the Japanese, but as moderately *unknown* risks to the Americans.

To further explore these differences we computed the mean nuclear risk scores for each cultural sample on the three dimensions constituting the *unknown* risk factor. In the U.S. sample, nuclear risks generally fell on the neutral to unknown side of the scale both for individuals and for science. They were intermediate on the new-old dimension. In sharp contrast, the same seven nuclear risks were perceived to be some of the most well known risks to the Japanese: nuclear risks were perceived as older and more well known to individuals and to science. The average individual knowledge score for the seven nuclear risks was 2.39 for the Japanese sample and

Table 4. **COMPARISON OF MEANS AND STANDARD DEVIATIONS FOR THE SEVEN RISK DIMENSIONS AND PROPORTION OF SIGNIFICANT T-TEST COMPARISONS: JAPAN AND THE UNITED STATES**

	Means and Standard Deviations[a]		Number of Significant t-Test Comparisons[b]	
Dimension[c]	Japan	U.S.	Japan > U.S.	U.S.> Japan
New/Old	2.23 (1.92)	2.54 (1.89)	14	33[d]
Unknown to science	3.21 (1.91)	2.85 (1.66)	35	21[d]
Unknown to individual	3.77 (2.04)	3.84 (1.83)	29	29[d]
Involuntary	4.39 (2.06)	3.79 (2.16)	38	10
Uncontrollable	3.46 (1.99)	2.74 (2.00)	51	2
Dread/Calm	4.35 (2.06)	3.81 (1.92)	43	5
Catastrophic	4.22 (2.18)	3.99 (2.06)	26	9

[a] Standard deviations in parentheses.
[b] Out of 70;p>.05.
[c] For each dimension higher scores mean, respectively, newer, unknown to science, unknown to exposed individuals, involuntary, controllable, dread, and catastrophic.
[d] All seven nuclear risks included.

4.50 for the American sample (here, a larger number means "newer," more "unknown to the individual," and more "unknown to science"). The old-new average score for the Japanese sample was 1.96, while for the U.S. sample it was 3.9. Likewise, the average scientific knowledge score was 2.05 for the Japanese and 4.01. for the Americans. All 21 t-test comparisons for the three dimensions were significant.

A score of 4 is the neutral point on a 7-point Likert scale. Thus, in all cases the average Japanese score was well below 4, while the average American score was either very close to or above 4. Thus, in both the relative comparison between the two cultures and in the Likert-type scale meaning comparison, Japanese students perceived nuclear risks as quite well known, while their American counterparts perceived them as slightly unknown or in between.

Examination of the dimensions - dread, controllability, voluntariness, catastrophic - associated with the *dread* risk factor revealed a mixed pattern of results for the seven nuclear items. On the dread dimension literally all nuclear risks were more dreaded in Japan than the United States. However, the overall relative difference was small (United States = 6.2 and Japan = 6.5). At the same time the 6+ scores for both cultures (on a 7-point Likert scale where the neutral point is 4) means that nuclear risks are highly dreaded in both cultures. On the other three dimensions constituting the dread factor - controllability, voluntariness, catastrophic - only a small fraction of t-test comparisons between the two samples was significant. Again, however, in both cultures the seven nuclear risks were viewed as highly uncontrollable, highly involuntary, and highly catastrophic.

Figure 1. **COGNITIVE MAP – UNITED STATES DATA**

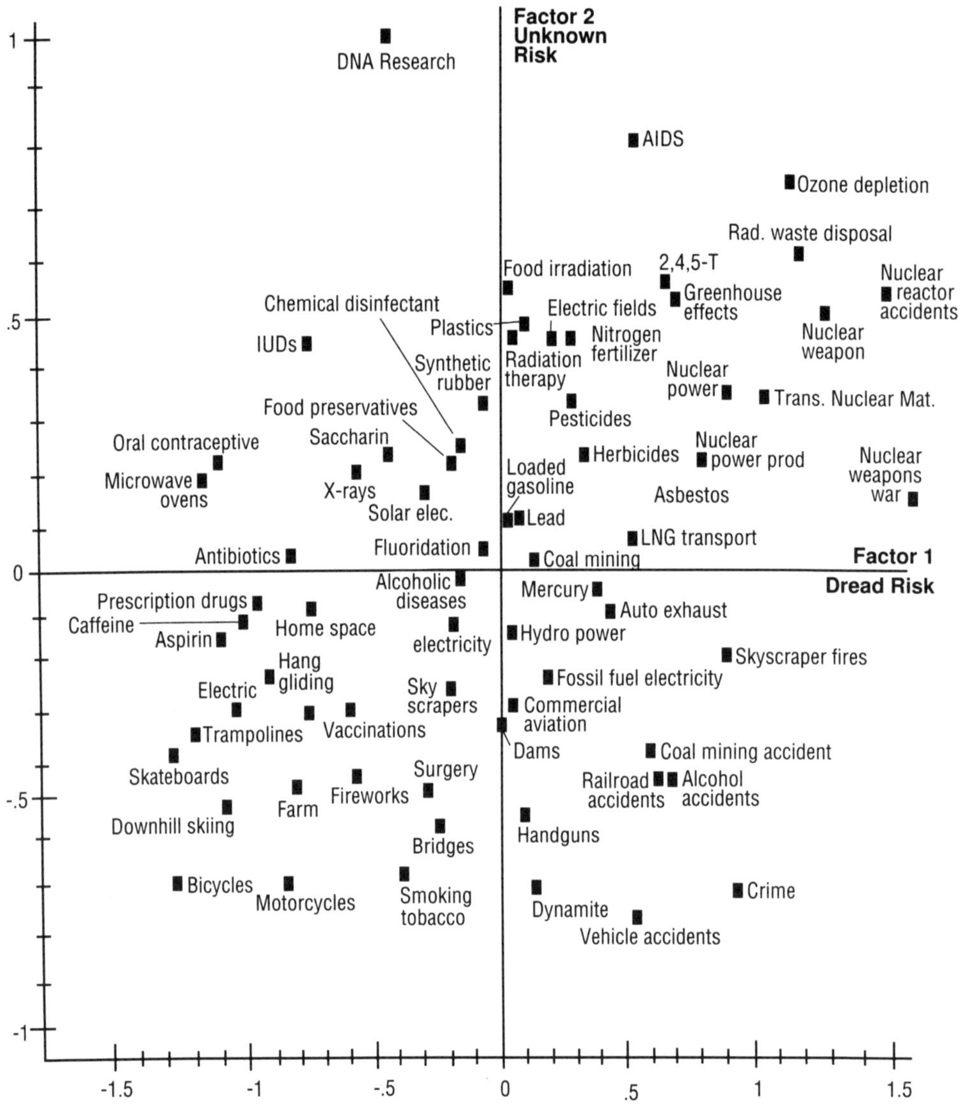

As with the pre-test and as with other studies within the psychometric paradigm we followed the factor analytic procedures of Slovic et al. (1985). The factor analysis, consistent with all previous research, resulted in two factors: *dread and knowledge.* The two factors explain between 60 and 65 percent of the variance. Then, again consistent with established procedures, factor scores were computed for each of the 70 risks by first weighting the numerical ratings on each risk

Figure 2. **COGNITIVE MAP – JAPAN DATA**

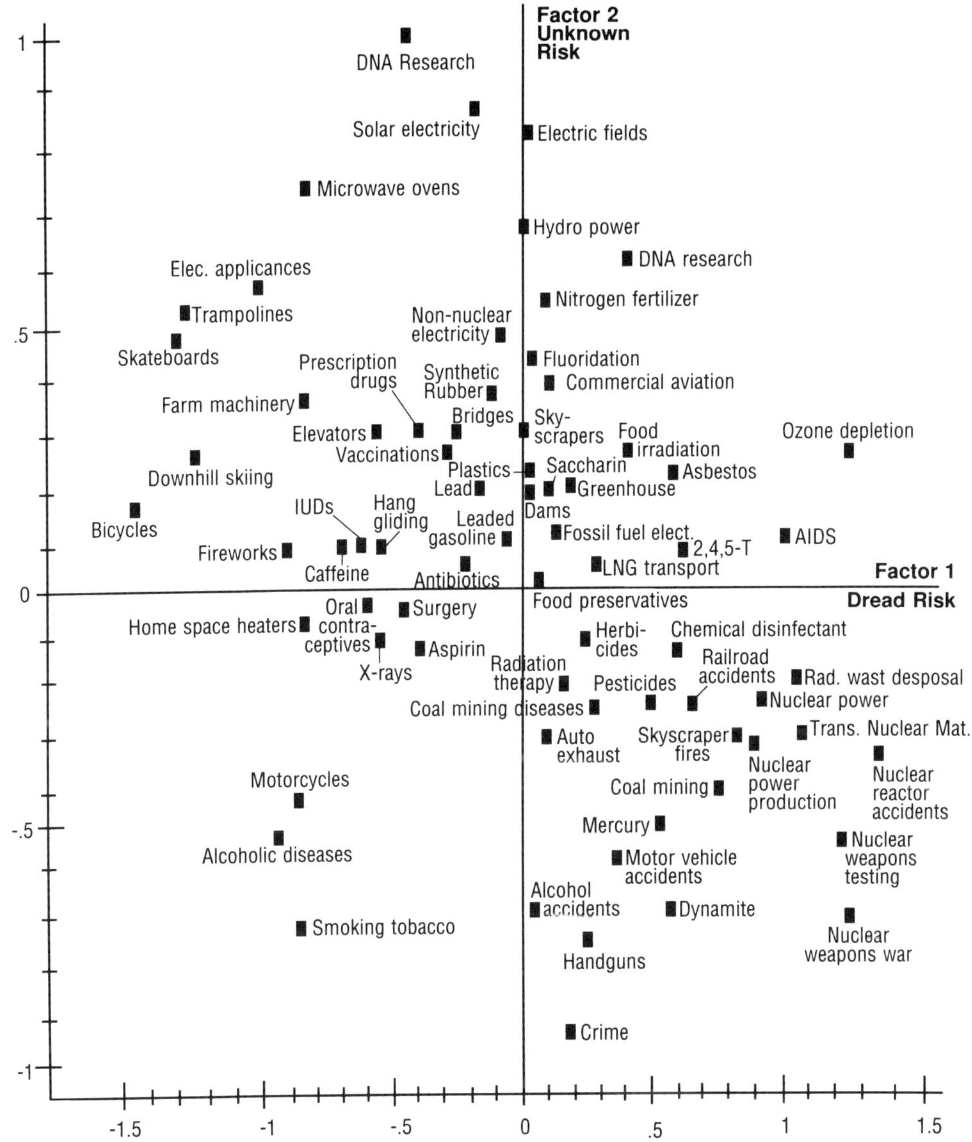

dimension proportionately to the importance of the dimension for the factor and by then summing over all dimensions. This weighted summation gives each of the individual 70 risks either a high or low score depending on whether the ratings are high or low on the dimensions most closely associated with each of the two factors. These results were then mapped for each sample separately into a two dimensional space. These cognitive maps are presented in Figures 1 and 2.

The *dread* risk axis comprises the four dimensions that loaded on this factor: dread-calm, uncontrollable-controllable, involuntary-voluntary, and catastrophic-not catastrophic. The *unknown* axis comprises the three dimensions that loaded on this factor: new-old, unknown-known to science, and unknown-know to individuals.

Perhaps the most compelling feature of the data for both samples is the clustering of the seven nuclear items. All nuclear items cluster together at the high dread end of the horizontal, dread axis. For Americans, nuclear war is particularly dreaded, a finding consistent with previous research (Slovic, Fischhoff, and Lichtenstein, 1985). The Japanese most dreaded a nuclear reactor accident. What is unmistakably clear is that nuclear risks evoke strong perceptions of dread in both cultures. At the same time, the cultures differ on the *unknown* risk factor - the vertical axis. All seven nuclear items are located on the upper section of that axis (the upper right quadrant) for the American sample, but on the lower section of that axis (the lower right quadrant) for the Japanese sample. As we found in the comparisons of means and in the t-tests performed above, the nuclear risks were perceived as somewhat unknown to Americans, but from somewhat to quite well known to Japanese.

This finding warrants elaboration. That the Japanese perceived nuclear risks as old and better known and that Americans perceived them as newer and more unknown makes sense in an historical context. Japan is the only society to have had nuclear weapons used against them in war and is, therefore, the only society to have direct experience with not only their explosive destruction, but also the massive doses of radiation they produced. The Nagasaki and Hiroshima bombings of August 1945 are a remembered wartime legacy in Japan. It is likely that this memory inclined the Japanese students to view all nuclear risks as older and better known than the American students.

Just as in the pre-test results of Phase I we find here that the Japanese students rated many of the medical (X-rays, oral contraceptives) and chemical (herbicides, pesticides) risks as significantly older and more known than did the American Students. And again replicating the pre-test results , the Japanese students perceived a variety of voluntary risks (bicycles, downhill skiing, hang gliding, fireworks, skateboards, trampolines, electric appliances, etc.) as less known than did their American counterparts.

DISCUSSION

The chief objectives of Phase II of the WSU Research Program were to administer concurrently, to samples of Americans and Japanese, the instrument developed in Phase I according to the protocols refined in Phase I. Both objectives were successfully attained. This resulted in a putatively valid and rich data set for subjecting the hypotheses underlying the entire research program to stringent tests.

The first set of tests with this data set, presented above, yielded results that are interesting, on the one hand, and challenging on the other hand. With respect to the cognitive maps, we found it

[3] Also referred to as covariance structure analyses or LISREL (the acronym for the first computer program—Linear Structural Relations Model)—written to develop and test such models.

quite interesting that they were dimensionally equivalent: they represented a two factor solution for both Americans and Japanese and explained similar amounts of variance for both. This provides some basis for arguing that the structure of risk perceptions in both cultures is, if not the same, at least quite similar. At the same time, we found differences in the positioning of certain clusters of risk items within the maps - the nuclear items being the most obvious of these.

A cognitive map can be viewed as analogous to a geographic map. In the latter case, if we asked an individual to draw a map of their country, we would expect that most people could produce some facsimile of it. We would further expect variability in the accuracy of the elicited maps, but most people should be capable of providing some pictorial representation of their country. Presumably, the representation would reflect the picture of their country's geography stored in their memory. Similarly, we would expect that the cognitive maps produced by individuals, with the aid of the mathematics of scaling procedures, are facsimiles of the cognitive landscapes of those individuals. Maps created from specific elements within those landscapes, such as risk perceptions, should likewise be pictorial representations of that domain of the landscape.

Tying the cognitive maps produced by our data back to their analogy, geographical maps, provides a framework for interpreting our results. We found that the contours of the cognitive maps between Americans and Japanese are similar, but with discrepancies in content. This resembles a situation where, after being asked to draw a map of their country, individuals produced general shapes that were more or less correct, but where specific landmarks - rivers, mountains, boundaries, etc. - were misplaced. Perhaps this is what we are observing in the cognitive maps of Americans and Japanese; they cognitively organize the collection of risks they are exposed within a space whose contours are similar, but where the placement of items in the cognitive space is quite different.

These findings return us to the hypotheses underlying the research program. Because of the similarity in the contours of the American and Japanese cognitive maps, can we now conclude that the risk perceptions of Japanese and Americas are sufficiently similar to support the pancultural continuity expectation? Or, is the distribution of the individual risk items within the maps sufficiently different that we are led to conclude that culture exercises a significant shaping influence on risk perceptions? In our view the results presented do not permit an unequivocal answer to these questions. The analytic techniques used here, as elsewhere in the psychometric paradigm, are insufficiently refined to permit an answer to such a fine-tuned question. In view of this, we argue for the need for analytic tools better suited to the tasks generated by these hypotheses. One such tool, we further argue, is multivariate causal modeling.

3.4 Phase III: Modeling

The specific tools suited to the task at hand are latent variable models.[3] Phase III or the WSU Research Program comprises the development and testing of a variety of latent variable models with the Phase II data. See Table 1.

ANALYTIC STRATEGY

It was our original intention to subject both the American and Japanese samples, to a single, general model and then to perform statistical tests to determine: (i) if the general models of the data, one for each culture, were similar or different, and (ii) regardless of what was found in (i), whether there were statistical differences on any of the seven dimensions underlying the data structure. We found this intention to be intractable on computational and other grounds and, therefore, abandoned that goal.

Instead, we have subjected sub-sets of the risk items to our modeling efforts. This effort is still in progress. We first modeled the sub-set of seven nuclear items (See Table 2) because of the interesting patterns they displayed in the cognitive maps between Americans and Japanese, and because they were the most dreaded items for both samples.

BASIC MODEL

In a typical factor analysis within the psychometric tradition, factor scores (as above) are typically obtained by computing weighted averages for each risk dimension. This is done to reduce the data structure from three to two dimensions so that factor analyses can be performed. In our modeling efforts we took a slightly different tack. In particular we performed a factor analysis, for the nuclear items, on the individual risk items and found that they loaded on certain of the risk dimensions. In other words, the measured perceptions of the individual risk items are the indicators of the risk dimensions - which in modeling language are called "latent constructs".

Figure 3. **BASIC LATENT VARIABLE MODEL**

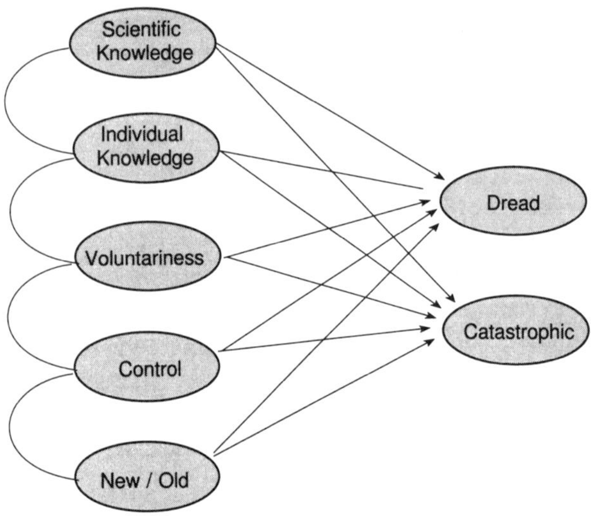

The factor loadings (343 in total) are not shown here to conserve space.[4] The pattern of loadings first revealed that the seven risk dimensions in our data set were independent constructs. For example, the *dread and catastrophic* dimensions were not part of a common construct,

[4] They are available from either of the first two authors.

[5] Goodness-of-Fit-Index provided by the program. It measures the squared discrepancies between the observed covariance matrix and the matrix generated by the model.

[6] An Adjusted-Goodness-of-Fit-Index. This adjusts the GFI by a ratio of the degrees of freedom of the model being tested.

dread, as is the interpretation in the psychometric tradition, but were separate constructs. Furthermore, the results suggested that they affect each other and, in turn, are affected by the other dimensions, or constructs. The path diagrams of these results are presented as Figure 3, which we label the "Basic Model".

We performed goodness-of-fit tests for this model using the EQS program (Bentler, 1989). In particular, we subjected the data to the Likelihood Ratio χ^2, the GFI[5] and AGFI[6] goodness-of-fit tests. We found that the basic model performed poorly for all indexes, regardless of sample. We, therefore, sought to refine the basic model.

REFINED MODELS

We accomplished our refinement by iteratively fitting a number of alternative models, with a reduced number of constructs, to the data. Small factor loadings or small path coefficients were the bases for dropping constructs. The result of this iterative process was a refined model for each data set, American and Japanese, that we have labeled "Best-fit" model. Figure 4 presents the Best-fit path model for each data set. The model coefficients are standardized path coefficients.

The Likelihood-Ratio χ^2 provided a poor fit for each sample: Japan $\chi^2_{(448)} = 822.629$, $p< 0.001$; United States $\chi^2_{(158)} = 222.395$, $p < 0.001$. However, the GFI and AGFI provided acceptable fits for both samples: Japan GFI = .807, AGFI = .772; United States GFI = .844, AGFI = .793. The GRI and AGFI indices indicate that the models perform fairly well in reproducing the original data matrices. On this statistical basis we did not reject these two refined models.

Figure 4. **BEST-FIT MODELS**

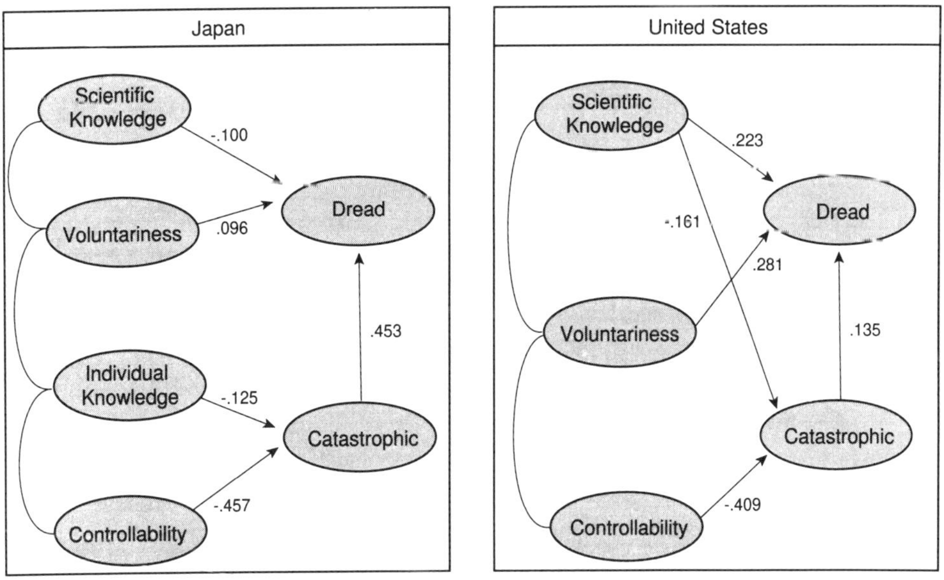

Descriptively, the structure of the models displays similarity and differences. The key similarity, corroborating the factor analyses with the basic model, is that the *dread* and *catastrophic* dimensions are separate constructs and that the catastrophic has a recursive influence on dread. Another similarity is that both of these constructs are recursively influenced by the other constructs in the model. The key difference is that the individual knowledge construct is retained in the Japan model but not in the American model.

Within a similarity of structure, the models represent a fundamentally different set of processes. For the Japanese, catastrophic images are a function of lack of individual knowledge about risks and having little control over them. In turn, dreaded images are a function of catastrophic images (and indirectly, individual knowledge and controllability), lack of scientific knowledge, and involuntariness. In contrast, the catastrophic images of Americans are a function of lack of scientific knowledge about the risks and, like the Japanese, lack of control over them. The dreaded images of Americans are, then, a function of catastrophic images (and indirectly scientific knowledge and controllability), the availability of scientific knowledge, and involuntariness.

We consider these results somewhat preliminary, so do not want to overdraw conclusions from them. Nevertheless, what we seem to be uncovering - at least for this subset of nuclear items - is that Americans and Japanese form similar perceptions about the dread and catastrophic potential of the risks, but arrive at that common point through different cognitive processes. The particular constructs and their directionality appear to be weighted differently in the two cultures. Our next task is to determine the robustness of these refined models. That task, currently in progress, involves fitting the models to other organized subsets of risk in the Phase II data set. Should the models hold up well when fitted to other sub-sets of our risk data it will not only confirm our interpretations, above, but also provide a firm foundation for claiming that we have uncovered the bases of similarities and differences in fundamental perceptual processes.

4 CONCLUSIONS

In this chapter we have provided a summary of the WSU Research Program still in progress. The program was designed to make systematic progress in understanding some of the basic human processes underlying perceptions of a wide variety of risks. In particular, the program was designed to demarcate between pancultural and culturally variant features of the cognitive processes of risks. One key element of the program was the careful selection of a comparative culture. For reasons argued above comparison between Japan and a Western culture provides one of the most stringent tests of the hypotheses guiding the research program. For convenience alone, our choice of a Western culture was the United States. Another key element was the development of a comparative instrument, in Phase I, that met a number of language translation criteria and tests of face validity.

Phase II of the research program produced a rich data set with which to test, not only the specified hypothesis, but also to perform sub-analyses and other refinements. This data set was first analyzed using the standard factor-analytic techniques of the standard psychometric paradigm. In general, the findings from the United States-Japan comparison replicated previous results produced within this paradigm - including the cross-cultural results. The cognitive maps produced by this analysis indicated common contours between the Americans and Japanese, but different cognitive content. Without a formal basis for judging whether contour or content should be the criterion for determining which of our hypotheses were supported, the fundamental questions generating the research program remained unresolved.

To resolve them we turned to latent variable modeling in Phase III of the research program, still in progress. Thus far this phase has produced preliminary results on only the sub-set of seven nuclear risks contained in the data set These results are, nevertheless, promising. We developed a refined, best-fitting model for each culture based upon the nuclear items. What these models seem to be telling us is that Americans and Japanese have similar images of dread and catastrophic potential underlying their perceptions of nuclear risks. But, importantly, the cognitive routes they use to arrive at those images are fundamentally different. This is intriguing. We are now in the process of testing the robustness of these models with other risks in our data set. We are confident that the results of this effort will be equally as intriguing.

In sum, though our research program is still ongoing, we believe we have made considerable progress in developing a richer understanding of the structure and processes by which humans form their perceptions of risks. Should that progress continue in our future work our research program will have more than fulfilled its objectives.

ACKNOWLEDGMENTS:

Portions of this work was funded by the Commemorative Association for the Japan World Exposition, by the Japan-United States Friendship Commission, and by the Edward R. Meyer Distinguished Professorship in Natural Resource and Environmental Policy in the Thomas S. Foley Institute for Public Policy and Public Service at Washington State University.

The authors express their appreciation to Yuki Fujioka, Sachiko Honda, and Michiko Joerding, and Monoru Chinen for performing the Japanese translations, to George Hinman, Thomas Loweinger, and Paul Slovic for assistance and support, to Rodney Baxter for data analyses, and to Tom Dietz, Rose Krebill-Prather, Kanehiro Hoshino, Ray Jussaume, Kazuaki Matsui, Allan Mazur, Riley Dunlap, Ortwin Renn, Lennart Sjöberg, and Brit Marie Drottz-Sjöberg for comments and criticisms at various stages of the research program described here.

5 REFERENCES

Bandura, A. 1986. Social foundations of thought and action: A social cognitive theory. Englewood Cliffs: Prentice Hall.

Bentler, P. 1989. EQS: Structural Equations Program Manual. Los Angeles, CA: BMDP Statistical Software, Inc.

Boholm, A. 1996. The Many Faces of Cultural Theory of Risk: Anthropological Misgivings. Paper presented at the annual meetings of the Society for Risk Analysis-Europe, Guildford, University of Surrey, UK.

Boholm, A. 1998. Comparative Studies of Risk Perception: A Review of Twenty Years of Research. Journal of Risk Research 1:135-163.

Brislin, R. 1993. Understanding Culture's Influence on Behavior. Fort Worth, TX: Harcourt Brace Publishers.

Byrne, B. M. 1994. Structural Equation Modeling with EQS and EQS/Windows. Thousand Oaks, CA: Sage.

Covello, V. T. and Kawamura, K. 1988. Cooperation vs. confrontation: A comparison of approaches to environmental risk management in Japan and the United States. Risk Analysis 8:161-175.

Covello, V. T. von Winterfeldt, D. and Slovic, P. 1986. Risk communication: A review of the literature. Risk Abstracts 3: 71-182.

Cvetkovich, C. and Earle, T. C. 1985. Classifying hazardous events. Journal of Environmental Psychology 13:5-35.

Cvetkovich, G., Vlek, C. and Earle, T. C. 1988. Designing public hazard communication programs about large-scale technologies, in C. Vlek and G. Cvetkovich (Eds.), Social Decision Methods for Large-Scale Technologies. Amsterdam: North Holland.

Deaux, K. and Wrightsman, L. S. 1988. Social Psychology, 5th edition. Pacific Grove: Brooks/ Cole.

Douglas, M. and Wildavsky, A. 1982. Risk and Culture. Berkeley: University of California Press.

Englander,T., Farago, K., Slovic, P, and Fischhoff, B. 1986. A comparative analysis of risk perception in Hungary and the United States. Social Behaviour 1:55-66.

Form, W. 1979. Comparative industrial sociology and the convergence hypothesis. Annual Review of Sociology 5:1-25.

Frane, J. W. and Hill, M. 1976. Factor analysis as a tool for data analysis. Communication Statistics A5(6):487-506.

Goszczynzka, M., Tyszka, T. and Slovic, P. 1991. Risk perception in Poland: A comparison with three other countries. Journal of Behavioral Decision Making 4:179-193.

Heisenberg, W. 1958. The Physicist's Conception of Nature. Trans. by Arnold J. Pomrtrans. New York: Harcourt Brace.

Hinman, G. W., Rosa, E. A., Kleinhesselink, R.R. and Lowinger, T.C. 1993. Perceptions of Nuclear and Other Risks in Japan and the United States. Risk Analysis 13:449-455.

Hoshino, K. 1989. Personal communication, National Institute of Police Science, Tokyo, Japan.

Johnson, E. J. and Tversky, A. 1984. Dimensional analysis of risk perceptions. Journal of Exuerimental Psychology:General 113:53-64.

Kahneman, D., Slovic, P and Tversky, A. (Eds.) 1982. Judgment Under Uncertainity: Heuristics and Biases. Cambridge: Cambridge University Press.

Karpowicz-Lazreg, C. and Mullet, E. 1993. Societal risk as seen by the French public. Risk Analysis 13:253-258.

Keown, C. F. (1989). Risk perceptions of Hong Kongese vs. Americans. Risk Analysis 9:401-405.

Kleinhesselink, R. R. and Rosa, E. A. 1994. Nuclear Trees in a Forest of Hazards: A Comparison of Risk Perceptions between American and Japanese University Students. Pp. 101-119 in Thomas C. Lowinger and George W. Hinman (eds.), Nuclear Power at the Crossroads: Challenges and Prospects for the Twenty-First Century. Boulder, CO: International Research Center for Energy and Economic Development, University of Colorado.

Kleinhesselink, R. R. and Rosa, E. A. 1991. Cognitive Representation of Risk Perceptions: A Comparison of Japan and the United States. Journal of Cross-Cultural Psychology 22:11-28.

Kraus, N.H. and Slovic, P. 1988. Taxonomic Analysis of Perceived Risk: Modeling Individual and Group Perceptions Within Homogeneous Hazard Domains. Risk Analysis 8:435-455.

Lichtenstein, S., Slovic, P., Fischhoff, B., Layman, M. and Combs, B. 1978. Judged frequency of lethal events. Journal of Experimental Psychology: Human Learning and Memory 4:551-578.

Loehlin, J. C. 1992. Latent Variable Models: An Introduction to Factor, Path, and Structural Analysis. (2nd ed.). Hillsdale, NJ: Lawrence Erlbaum Associates.

Mazur, A. 1982. The Dynamics of Technical Controversy. Washington DC: Communications Press.

National Research Council. 1989. Improving Risk Communication. Washington DC: National Academy Press.

Nyland, L. G. 1993 Risk Perception in Brazil and Sweden. Rhizikon Risk Research Reports No. 15, Stockholm: Center for Risk Research, Stockholm School of Economics.

Renn, O. and Swaton, E. 1984. Psychological and Sociological Approaches to the Study of Risk Perception. Environment International 10:557-575.

Rosa, E. A. 1998. Metatheoretical Foundations for Post-Normal Risk. Journal of Risk Research 1:15-44.

Rosa, E. A. and Kleinhesselink, R.R. 1990. Risk Perceptions in Japan and the United States: An Empirical Test of Cultural Influences. Paper presented at the annual meetings of the Pacific Sociological Association, Spokane, WA, USA (April).

Rosa, E.A. and Matsuda, N. 1997. The Cognitive Architecture of Risk Perceptions: Pancultural Unity or Cultural Variation? Paper presented at the annual meetings of the Society of Risk Analysis, Washington, DC, USA (December).

Sears, D. 0. 1986. College sophomores in the laboratory: Influences of a narrow data base on social psychology's view of human nature. Journal of Personality and Social Psychology 51:515-530.

Sjöberg, L. 1995. Explaining risk Perception: an Empirical and Quantitative Evaluation of Cultural Theory. Rhizikon Risk Research Reports No. 22, Stockholm: Center for Risk Research, Stockholm School of Economics.

Slovic, P. 1987. Perception of risk. Science 236:280-285.

Slovic, P. 1986. Informing and educating the public about risk. Risk Analysis 6:403-415.

Slovic, P., Fischhoff, B. and Lichtenstein, S., 1985. Characterizing perceived risk, Pp. 91-125 in R. Kates, C. Hohenemser, and R. Kasperson (Eds.), Perilous Progress: Managing the Hazards of Technology. Boulder CO: Westview Press.

Teigen, K. II., Brun, W. and Slovic, P. 1988. Societal risks as seen by a Norwegian public. Journal of Behavioral Decision Making 1:111-130.

Thompson, M., Ellis, R. and Wildavsky, A. 1990. Cultural Theory. Boulder, CO: Westview Press.

6

CROSS-CULTURAL RISK PERCEPTION:

STATE AND CHALLENGES

Ortwin Renn
Center of Technology Assessment, Stuttgart, Germany

Bernd Rohrmann
University of Melbourne, Melbourne, Australia

TABLE OF CONTENTS

SUMMARY

The perception of risk within different cultures is a rather complex phenomenon that cannot be described on the basis of a single theory or model. The major accomplishment gained through psychological research was the discovery of qualitative risk characteristics and cognitive as well as affective factors that serve as heuristic tools for classifying and evaluating risk sources or activities. Sociological and cultural research has been essential in pointing out the social, political, and cultural factors that influence risk perception directly, as well as indirectly through the assignment of significance to the individual risk characteristics. The relative weights of the individual risk characteristics depend on social and cultural factors that form the main research agenda for cross-cultural research. The studies compiled in this volume include trust and credibility, social values, and group affiliations as predictors for risk responses. This last chapter provides a review of the findings of these studies and outlines the main lessons for understanding risk perception and designing effective programs for risk management and communication, as well as for improving future cross-cultural risk research.

1 INTRODUCTION

Over the last decade, risk perception research has changed from focussing on a single culture to a cross-cultural perspective. This shift has been fostered by several studies devoted to the investigation of cultural and national differences and commonalties between samples of different countries. The contribution by Rosa et al. in this volume started with two alternative hypotheses about the nature of risk perception. The first proposition was to assume that risks are expected to be perceived in the same basic way across cultures because humans -as a common species- have a common cognitive architecture that filters perceptions in the same way. The second proposition was to assume that human perceptions (including perceptions of risk or uncertainty) are so embedded within cultural meaning that they are expected to exhibit the same variability as other aspects of human experience.

In addition to the allegedly universal cognitive processes that govern risk perception in all or at least in most social and cultural contexts, our focus has been on culture-specific mechanisms of processing and evaluating risks. We have also addressed a variety of factors hypothesized to exert substantial influence on risk perception (see contributions by Slovic et al., Sjöberg et al. and Rohrmann). Among such factors were group identity, economic status, social and personal values, professional orientation, and others. Which of these factors influence risk perception in what way? Have these empirical investigators been able to identify universal characteristics for risk perception? And if so, how do they interact with other personal, social and cultural influences? How certain are we that correlation found in one social or cultural context can be assumed to be valid in other similar or even different contexts?

The contributions in this volume do not give conclusive, let alone, final answers to these questions, but they provide some compelling evidence about the nature of risk perception seen through the lenses of people from different countries and different social backgrounds. The most obvious result is that the picture of risk perception drawn from comparative analysis is more complex than many theorists have assumed so far. The simple juxtaposition of universal heuristics on one hand and cultural prototypes on the other hand needs replacement by a more refined model of risk perception that includes anthropological, psychological, social and cultural variables. Furthermore, these different levels of influence are not independent of each other but form a web of interrelations that defy any attempt to reduce risk perception to a single set of variables.

The last chapter of this book has three objectives: Firstly, it summarizes the main results of the studies in this volume from a theoretical as well as methodological point of view. Secondly, it suggests a framework for understanding and investigating risk perception using a model of five context levels of influencing factors. Thirdly, this chapter includes some recommendations for the use of risk perception studies in policy making and for further cross-cultural research.

2 RECENT FINDINGS FROM CROSS-NATIONAL STUDIES

The four studies described in this book differ in their choice of countries, in the research question(s) on which they are focused, and in the selection of methods employed (sampling approach, questionnaire constructions and data analyses). Table 1 provides an overview of the four studies. The table lists the sampling methods, the countries involved in each study, and the primary dependent and independent variables. It is interesting to note that all studies invest a major effort to define and differentiate the dependent variables. While earlier studies on risk perception mostly used a single scale for measuring perceived risk, more recent investigations and those represented in this book have been very careful in their attempt to understand the complexity of the perception process. The distinction between risk perception related toward oneself or one's relatives and friends, and risk perception related to society as a whole has become standard in the risk perception literature. Furthermore, perception of benefits in addition to risk characteristics, perceived magnitude, probability or severity, perceived threat or perceived health impacts - these are all different target variables when it comes to specifying the linguistic space associated with risk perception.

Among the independent variables that have been used in the four studies are the classic psychometric variables, ideological background and worldviews, including the four cultural prototypes suggested by Douglas and others (Douglas 1985; Rayner 1984, 1987): affections or anxieties, socioeconomic status, professional affiliations, trust and others. Similar to the treatment of the dependent variable, the field of independent variables has become more sophisticated over the last years. At the same time, however, the field has made some progress in refuting simple models of risk perception. The wealth of knowledge gained from the studies in this book and from other recent investigations has demonstrated clearly, that neither the claim of irrationality

Table 1: **OVERVIEW OF THE RESEARCH PROJECTS REPORTED IN THIS VOLUME**

Study	Sampling	Countries	Dependent var.'s	Independent var.'s
Slovic et al.	Random general population	USA, France	Magnitude of risk, benefits (personal and social), acceptability	Qualitative hazard characteristics, cultural prototypes, levels of affect, demographics, country
Rohrmann Study I	Special groups (ecological, technological, monetarian, feminist orientation)	Germany, Australia, New Zealand	Risk appraisals, incl. magnitude, health impacts, catastrophic potential, benefits, risk acceptance (personal and societal)	Cognitive beliefs, other attitudes, group composition, country
Rohrmann Study II	Students (three different disciplines), scientists	Australia, China	Same as study I	Same as study I
Sjöberg et al.	Special Groups (Nurses, Students of economics & business, skilled workers, homeless people)	Bulgaria, Romania, Sweden (Brazil as contrast)	Magnitude (personal; general); demand for personal and social mitigation, and others	demographic variables & gender, respondent group, anxiety, credibility of risk information sources, country
Rosa et. al.	Students of business and psychology	USA, Japan	Hazard characteristics, incl. dread, familiarity, control	Sample features, country

as the main source of intuitive risk perception, nor the claim of a deterministic relationship between universal risk characteristics and perceived risk can be sustained in the light of empirical evidence presented. Before we try to draw our own conclusions from this wealth of empirical material, it is necessary to summarize first the main findings of each study.

The first study, by *Slovic et al.*, is directed toward a better understanding of people's attitudes towards nuclear power. The comparison was done for two countries: the United States and France. The samples in both countries showed many similarities but also noticeable differences. Among the similarities two aspects are specifically noteworthy.

First, in contrast to public belief, the French respondents were just as concerned about the hazardous nature of nuclear energy as the American respondents. In spite of the fact that nuclear energy is still promoted in France and politically welcomed by most parties and institutions, most French respondents rated nuclear energy as well as nuclear waste as belonging to the highest risk type. In both countries, concerns focused on issues of safety, health, and environmental quality. Even though perception of risk proved to be a strong predictor of attitudes toward nuclear power within both the U.S. and France, it cannot account for the different level of reliance on nuclear energy in the two countries. The structure as well as the directions of each country's attitudes turned out to be rather similar.

Second, women in both countries were more critical about nuclear power than men. Further analysis revealed that around 30% of white males (in the United States) and a slightly larger male group in France stood apart from all other respondents in having low perceptions of risk from all hazards and showing strong support for nuclear power. In contrast to this finding, the authors found that the subgroup of black respondents from the U.S.-sample did not include a substantial number of risk-prone males. Strong supporters of nuclear energy consisted predominantly of white males. Slovic at al. concluded that this group of strong supporters of nuclear energy form a "white male subculture" of risk-prone, technology-embracing and economic growth endorsing individuals who believe in the merits of scientific progress and risk-taking behavior as a source of wealth. They tend to have higher incomes and education and hold highly authoritarian and anti-egalitarian attitudes. They also demonstrated more trust in experts than the average respondent. If one removed this group from the analysis, the differences between men and women almost vanished.

There were also clear differences between the two countries: the French respondents saw greater need for nuclear power and expected greater economic benefit from it. They had greater trust in scientists, industry, and government officials. The French respondents believed that politicians and experts should make the necessary decisions about nuclear power rather than the people affected by these decisions. The U.S.-respondents showed much more distrust in experts, were more inclined to support direct citizen participation, and opted for more democratic control over the decision making process. It became clear that the differences between the French and the U.S. public could not be traced back to differences in perception of risk or benefits but to differences in the evaluation and practice of the two political cultures. The majority of French respondents had confidence in the centralized political decision making bodies who rely on expert consultation and bureaucratic reasoning (Renn 1995). Legitimization in France involves the proof that the best experts have been consulted and that some agency is willing to accept the responsibility (and accountability) for a decision that deviates from what the public actually wants or prefers. This trust-relationship is alien to the U.S.-culture. People demand that their preferences are taken into account when decisions are being made. In addition, they do not trust agencies to be honest and independent brokers in managing public safety, health, and environmental affairs. Thus, citizens in the U.S. expect their public officials to reflect what the citizens perceive as dangerous and act according to the revealed preferences of their clients.

The study by Slovic et al. also tested the validity and applicability of the cultural prototype approach for explaining risk perception. The results showed that the four prototypes were not exclusive variables by which people could be identified as belonging to one or another prototype. Rather these prototypes resemble ideals that influence individuals to varying degrees. These ideals are indeed correlated with risk perception - in the U.S.-sample more strongly than in the French sample. The data indicate that those favoring a more egalitarian society, in which power and wealth are more evenly distributed, tend to be anti-nuclear. Those persons endorsing individualistic attitudes or attitudes depicting a hierarchical social system, tend to be pro-nuclear.

The importance of cultural variables for risk perception is also emphasized in the second study of this book, which investigates risk perceptions and attitudes in Germany, Australia, New

Zealand and China. *Rohrmann* did not use random samples but focused on societal and/or professional groups. In his first study he included people with a technological, monetarian, ecological or feminist orientation, while in the second one students and scientists from the fields of psychology, technology, and geography were surveyed. Although Rohrmann found considerable cross-national variations in risk perception among the countries as well as among the four groups, he warns the reader against using this data as an implicit proof for the validity of the cultural prototypes that Slovic at al. explicitly tested in their study. In spite of some similarities between the sample groups and the four prototypes, Rohrmann explains the differences between the four groups predominantly by referring to the impact of professional values and worldviews that govern the behavior of group members. It was indeed surprising that the differences between the countries were less pronounced with respect to most variables used in the first study (more pronounced differences appeared in the second study including China) than between the four groups within one country. This finding amplifies the observation that in the process of globalization national identities are partially exchanged for professional identities worldwide. One could even claim that the bankers or feminists around the world have more in common than the individual banker or feminist with their own children belonging to another cultural or professional group.

Rohrmann also pointed out that among the many hazards that people might worry about several are always perceived as more serious than others, regardless of country or professional group. The list of 'top' hazards included nuclear power, asbestos work, and smoking. There were also some noticeable differences in the rank order of hazards between countries. Most of the variance could be explained by the differences in exposure levels or significance assigned to these hazards in each country. However, it would be misleading to assume that risk perception mirrors the measured level of risk revealed through risk assessment. Rohrmann demonstrated that many hazards were perceived as either more perilous or less severe than epidemiological data would suggest. This finding was prevalent in each of the countries included in the analysis.

The data on China (*Rohrmann's* study II) differed most from the results found in the other three countries. The Chinese assigned particularly low risk acceptance scores to most hazards, at least regarding individual acceptance, less though for societal acceptability. As naive observers and travelers to China we would have suggested the opposite, given the fact that developing countries value risk-taking behavior as a means of promoting economic growth and that the Chinese culture is based on collectivism and individual sacrifice rather than individualism and self-realization. But experts on Chinese culture emphasize that risk-taking is especially repugnant to Chinese decision makers and that high risk options are likely to be rejected due to risk aversion among collective decision makers. One might even draw a line to the study of Slovic et al.: The French trust in public officials was accompanied with the belief that this trust is given in exchange for assuring low risk exposure. Something similar may be in effect in China: Strong respect for status and authority is linked with the conviction that these authorities will do their best to reduce the risks to society to a minimum. In spite of the differences in perceived levels of risk, the Chinese respondents rated risks not dramatically different from the rank order that was found in the other three countries, with the exception of some lifestyle risks, especially those that are socially banned (e.g. gambling or use of hallucinogenic drugs).

The fact that risks are ranked in a similar fashion was also found in the third study reported in this book. *Sjöberg et al.* compared six different groups in different national contexts. The main data were collected in Bulgaria and Romania. The results of these two surveys were also compared to results of studies conducted in Sweden and Brazil. The six groups were: nurses, workers, economists, engineers, teachers, and slum dwellers. Sjöberg et al. were able to show that the ranking of risks, both personal and general, were rather similar between the two countries, and with the exception of a few outliners between the six groups as well. The similarities in risk ranking could also be detected in the two comparative samples from Brazil and Sweden. People seem to be concerned about the same threats all over the world. The authors observed several exceptions, however, in the study of the two eastern countries, for example, the risk to be assaulted, to be struck by lightening and to be attacked by wild dogs in the street. These differences may reflect subjective expectations with respect to exposure in each country. Technological risks are rated almost identically in the two countries.

Another similarity to Rohrmann's study was the finding that the groups in Bulgaria were characterized by larger differences in risk perception scores compared to the differences between each national sample. The means of the Romanian groups were quite close to each other, however. In four cases, group specific perception patterns prevailed in the two countries: Nurses rated risks higher than on average, workers and economists were close to average, and engineers gave lower ratings than the other groups. This was true for both Bulgaria and Romania. The two nations differed on the two remaining groups. Teachers gave low ratings in Romania and high in Bulgaria. Slum dwellers gave high ratings in Romania and low in Bulgaria.

With respect to the independent variables, Sjöberg et al. reject the thesis that the four cultural prototypes are powerful predictors for risk perception. The results of their study suggest that the risk judgments made by the Bulgarian and Romanian samples are based on the perception of specific threats and perhaps media risk salience, rather than general culturally contingent values and beliefs. Cultural orientations and biases are important variables that influence risk perception but they may differ from those specific biases explicated by the cultural theory of prototypes. Furthermore, economic status and social aspirations may be more decisive in shaping people's attitudes towards technological risks than cultural biases.

The common observation that risk perception rankings are rather similar among different nations and cultures, and that some of the factors influencing these rankings are almost identical has inspired the study by *Rosa et al.*. This research paper completes the number of empirical studies reported in this book. The intention of this study was to test the hypothesis of "universalism". The authors included several student samples from the United States and Japan. The main objective of the study was not to find the best predictors for explaining variations in risk perception but to reveal hidden patterns of risk perception based on psychometric characteristics among and between the different samples.

A particularly interesting result of this study was that the main factors governing the risk perception process in the United States could also be found in the perception process of Japanese students. Based on previous studies by Slovic et al. (1985) about the importance of risk characteristics for perceived magnitude of risk, two of the most powerful risk characteristics, the dread factor and the catastrophic potential, were also confirmed as valid predictors for overall risk perception scores in both countries. But, even more importantly, the cognitive routes the respondents used to arrive to those images were fundamentally different. For the Japanese, catastrophic images are a function of lack of individual knowledge about risks and having little control over them. In turn, dreaded images are a function of catastrophic images, lack of scientific knowledge, and the involuntary nature of the phenomena. For the American respondents, these relationships are almost reversed. In particular, catastrophic potential is here a function of lack of scientific knowledge about risks, not individual knowledge. This result is surprising since most observers of Japanese society are convinced that collective knowledge enjoys high esteem there. If these results are confirmed they may suggest different routes for risk communication. Americans need to be convinced that science is able to limit the maximum credible loss, whereas the Japanese need to be assured that people can rely on personal experience in dealing with such risks and that personal knowledge is available to cope with the potential outcome of a hazardous event.

3 PRINCIPAL LESSONS FOR UNDERSTANDING RISK PERCEPTION

Although several decades of research have created a substantial base for understanding how people perceive and respond to risks, the investigation of risk perception still lacks a comprehensive conceptualization of the social experience of risk, i.e., the social processing of uncertainty and the perception and evaluation of expected consequences related to an event or activity. So far, the social science approaches to risks constitute a mixture of several different schools and perspectives.

Psychological research into risk perception has revealed that contextual factors shape individual risk estimations and evaluations (as summarized in chapter 1 of this volume). The identification of these factors, such as dread, the voluntary nature of the event, personal ability to control risks, familiarity with the hazard, associations of fear and the catastrophic potential, provides useful information about the elements that individuals process for constructing their interpretation of risks. These rather universal risk characteristics were also found to be influential in the studies by Slovic et al., Rohrmann and Rosa et al. in this volume. In addition, analyses of people's heuristics in making inferences have elucidated on how risk information is generalized and evaluated intuitively (Kahneman and Tversky 1974; 1979). These psychological studies fail to explain, however, why individuals select certain characteristics of risks and ignore others (Jasanoff 1998). Furthermore, focusing only on the individual as an information processor, these studies cannot fully grasp the societal and cultural variance of risk interpretations.

Sociological analysis provides some further insights into the social, cultural and organizational factors that influence risk perception (Shubik 1991; Clarke 1989). Some studies attempt to identify social influences in the formation and change of attitudes towards risk-bearing activities or technologies (Short 1984; 1989; Gould et al. 1988). Some aspects, such as perceived fairness in the distribution of risks and benefits, have gained special attention as part of the dynamic interaction among the various groups involved in rejecting or legitimizing a proposed imposition of a risk on a certain population (Kasperson and Kasperson 1983; MacLean 1986; Rayner and Cantor 1987). More theoretically oriented studies have emphasized the social construction of risk interpretations and their affinity to different types of knowledge acquisition, social interests, and cultural values (Bradbury 1989; Cvetkovich and Earle 1991, Kunreuther & Slovic 1996). Another major predictor for risk perception has been identified as trust in the scientific and political elites (Kasperson, Golding & Tuler 1992; Earle and Cvetkowich 1995). In this volume, Sjöberg et al. emphasized the role of social-economic factors, confidence in information sources and the influence of media coverage on the risk perception process. Both Sjöberg et al. and Rohrmann pointed out that professional standards, values and worldviews exert a major influence on risk perception which may even exceed the differences between distinct nations and cultures.

One contestant for a holistic cultural theory of risk is the proposal of cultural prototypes that are supposed to govern the risk perception process (Douglas and Wildavsky 1982; Rayner 1990; Schwarz and Thompson 1990). According to this approach, cultural beliefs and world-views determine how people experience and interpret risks. As stated in the introductory chapter, cultural groups have been classified into four generic types: entrepreneurial, egalitarian, bureaucratic, and stratified individualistic; some authors add "autonomous" as a fifth type. Each of these cultural types is supposed to develop its own criteria and selection rules for constructing a group-specific interpretation of risk. How people experience risk is thus a function of their cultural beliefs and values. This hypothesis has been partially supported by the study of Slovic et al. but was rejected by Sjöberg et al. in their contribution. Rohrmann suggested that this conceptualization may explain a specific portion of the variance in risk perception, but could not be taken as a universal predictor to understand risk perception in different cultural, social and economic contexts. The main problem with the concept of cultural prototypes (cf. also Marris et al. 1996) is insufficient validity; as many people cannot be classified according to the postulated types, the predictive power of this theory is hard to assess.

The array of individual and social factors that shape risk perception demonstrates first and primarily that the intuitive understanding of risk is a multidimensional concept and cannot be reduced to a quantitative algorithm such as the product of probabilities and consequences (Allen 1987). Although risk perceptions differ considerably among social and cultural groups, as it was clearly shown in the papers of this volume, there also appears to be an agreement about the principal nature of risk perception. Essentially, the meaning of risk is multi-faceted, and beliefs about the causes and circumstances of risk tend to be integrated into a consistent belief system which might guide or even "override" the evaluation of hazard-specific information.

Based on the results in this volume and the rich body of literature on risk perception reviewed in the first chapter of this book, we developed a structured framework of factors that might help analysts of risk perception to understand and interpret their empirical results (see Fig. 1). The figure illustrates the influences on risk perception related to four context levels: cultural background, social-political environment, cognitive-affective factors, and general heuristics. Each represents substructures of collective and individual influences. Each level is embedded in the next higher level to highlight the mutual contingencies among individual, social and cultural variables.

Figure 1. **FOUR CONTEXT LEVELS OF RISK PERCEPTION**

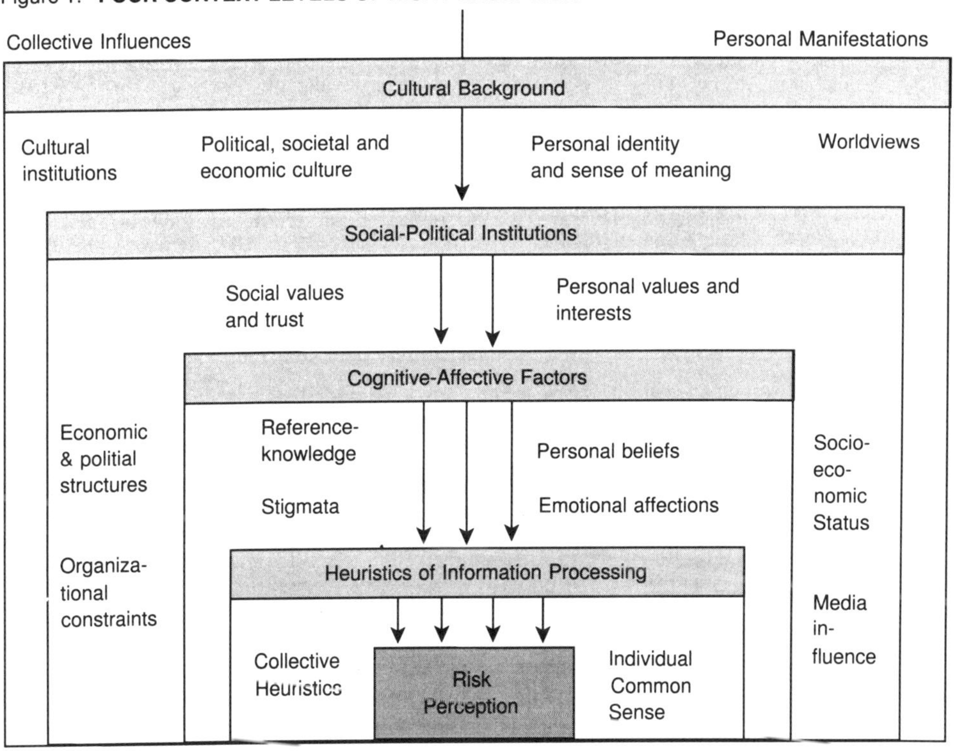

The first level includes the collective and individual heuristics that individuals and groups use when forming judgments about risks. Many researchers were able to demonstrate that several common characteristics such as dread, catastrophic potential, perceived controllability and familiarity have a major influence on the perceived magnitude of risks, based on empirical data from countries such as the United States, Canada, Germany, France, Austria, Japan and Australia (overview in Renn 1990, Rohrmann 1999). Images of dread, catastrophe and familiarity constitute tools of reducing complexity by providing easily identifiable cues for ordering new risks into prototypical cognitive schemata, such as the four images presented in the introductory chapter: pending danger, slow killers, cost-benefit ratio and recreational thrill. These images are internalized through cultural and social learning, and may be utilized for distinctions between negligible, serious, and unacceptable risks. The degree to which pertinent qualitative characteristics are assigned to specific risk sources depends on the cultural context, but also on the media coverage and the processing of secondary information as Sjöberg et al. were able to show in the comparison of Bulgarian and Romanian respondents. In fact, most people need to utilize secondary sources, if primary evidence is not mentally available for the hazard in question. The application of qualitative characteristics acts therefore as a primary mechanism of risk perception; its specific manifestation and relative importance depends, however, on the social and cultural context in which the individual is raised.

The second level refers to the cognitive and affective factors that influence the perception process directly and also indirectly through the assignment of special weights to universal heuristics. It has been common knowledge in perception research that cognition about a risk source, i.e. what people believe is true about a risk, govern the process of evaluation as well as the functioning and selection of universal characteristics. Interestingly Rosa et al. note that different cognitive processes could lead to the same result, in their case the arousal of catastrophic images. While the Japanese base their image of catastrophe on individual knowledge and familiarity, American respondents associated collective scientific experience and knowledge with catastrophic potential. This difference in cognitive routes also justifies the distinction between cognitive factors and risk characteristics (first and second level in Figure 1).

In contrast to cognitive factors, emotions have been widely neglected in risk perception research. Both Rohrmann and Sjöberg have demonstrated in their studies, however, that emotional feelings about risk sources influence risk appraisals and co-determine the salience of universal images such as dread and catastrophic potential. On the collective level, stigmata referring to risk sources or activities, play a similar role in stimulating emotional responses (Slovic et al. 1991). Recent work on risk perception of technological hazards also shows that emotional and cognitive factors are mutually related (Lantermann et al. 1992; Zwick and Renn 1998; Kals et al. 1999). It is not yet clear whether cognitive beliefs trigger the respective emotional responses or whether first emotional impulses "search" for arguments supporting one's emotional stance.

The third level refers to the social and political framework in which individuals and groups operate. Most studies on this level focus on the following influences: trust in institutions, personal and social value commitments, organizational constraints, social and political structures, and the socio-economic status of each individual. Trust, status, and values have been included in several of the comparative papers in this volume. An important variable in evaluating risk is the perception of fairness and justice in allocating benefits and risks to different individuals and social groups (Kals 1996; Linnerooth-Bayer and Fitzgerald 1996). Other studies have shown the relevance of political and social structures as well as organizational constraints (Clarke 1989; Short and Clarke 1992; Freudenburg 1992; 1993). Socio-political variables play a major role in shaping individual and social responses to "public" risk, and in structuring risk debates.

In a social environment in which personal experience is largely constructed by second-hand information, trust is an essential prerequisite for communication and social coordination (Luhmann 1990; Giddens 1990). In risk arenas, trust can easily be destroyed by non-predictable or non-avoidable disasters; at the same time, trust can easily be abused by referring to random events as explanations or excuses for risk management errors or hubris. As both processes occur simultaneously, trust is constantly at stake in institutional responses to risk. One way to cope with this problem is to develop an organizational style that either leads to high-reliability performance or to immunization against public scrutiny. In addition, the ambiguity in assigning causality or blame to various actors (including nature or God), makes risk an ideal issue for political maneuvering. Politically motivated actors in risk arenas can mobilize public support by placing blame and responsibility onto other actors. They know that the probabilistic nature of risk prevents any of the actors to demonstrate unequivocal evidence that their assessments are indeed correct and the competing assessments wrong. This ambiguity has two consequences. First, resources other than evidences, such as value commitments and solidarity, become major 'chips' in the game for more social influence in risk debates. Second, the position of regulators tends to become weak as uncertainty and ambiguity of risk issues make it increasingly difficult to act as neutral brokers of the common good (Renn 1992b). Whether this genuine feature of risk debates predominate the national debate in each country, depends on the political culture and the trust assigned to experts and authorities as Slovic et al. were able to show in their comparison of France and the United States. However, there is an almost universal tendency toward the erosion of trust even in traditional high-trust societies when it comes to high-risk technologies such as nuclear energy or genetic engineering.

In addition to trust, the media, social groups, institutions, and organizational bonds shape the individual and societal experience with risk. Several studies in this volume were specifically interested in the role of the media. Press media coverage appears to contribute substantially to a person's perception of risk even when controlling for the extent of direct harm to people and property. This finding matches the theoretical claim in the first chapter that risk reflects both, real harm and social constructions. In contrast to popular belief, however, there is no evidence that media create opinions about risks or even determine risk perceptions. Studies on media reception (cf. e.g. Coleman 1993; Lichtenberg and Mac Lean 1988; Peters 1986) suggest instead that people select elements from media reports and use their own frame of reference to create understanding and meaning. This creation is certainly influenced by the cognitive content and the evaluative statements of the reports, but the influence is more indirect and ambiguous than expected. Most people reconfirm existing attitudes when reading or viewing media reports (e.g., Peters 1994, Wilkins 1991).

In addition to media coverage, the papers in this volume addressed the issue of value commitments. These values were indirectly assessed through the selection of different groups in the studies by Rohrmann and Sjöberg et al. It was evident that differences in group affiliations were accompanied by major differences in risk perception (with the exception of Romania). Although group differences cannot be reduced to differences in values, it is obvious that these professional and group-specific values play a major part in constituting a group's identity. In addition, Slovic et al. were able to show that a subgroup of white males in the United States rated risk consistently low and expressed a high preference for authority, hierarchy, and economic performance. This again supports the notion that value commitments belong to the most influential factors in shaping risk perceptions. A series of recent studies on the perception of biotechnology (Urban

1998; Hampel and Pfenning 1998; Renn & Zwick 1997) provides further evidence for this interpretation.

The last level refers to cultural factors that govern or co-determine many of the lower levels of influence, including the selection of universal heuristics. The authors in this book expressed different opinions on the validity of the cultural theory of risk. Slovic et al. regard this approach as useful in explaining some of the differences in risk perception; Sjöberg et al. found the variance explained by cultural prototypes to be so low that they rejected the whole concept. Rohrmann expressed a skeptical view mainly because of measurement deficiencies. Rosa et al. did not comment on the validity of the cultural approach. All authors agreed however that specific culture-based preferences and biases are indeed important factors in risk perception; the disagreement is about the relevance of the postulated four prototypes within the realm of cultural factors.

All four levels of influencing factors are relevant for a better understanding of risk perception. The comparative studies have shown that to a certain degree universal yardsticks to evaluate risks exist but that their relevance and their selective use in assigning these characteristics to specific risk situations are contingent on cognitive and affective patterns as well as on the social, political and cultural environments. In spite of many remaining uncertainties and ambiguities in risk perception research, one conclusion is beyond any doubt: Abstracting the risk concept to a rigid formula and reducing it to probability and consequences violates people's intuitive feeling of what is important when making judgments about the acceptability of risks. Rather than evaluating risk with a single formula, most people use a set of multiple attributes, many of which make normative sense.

4 IMPLICATIONS FOR RISK MANAGEMENT

From this comparative review it has become evident that the social sciences are still struggling with risk perception. However, if approaches such as psychological, sociological, and cultural conceptualizations are looked at together rather than viewed as separate entities, they provide a wide range of theoretical insights and empirical knowledge that can help risk managers to understand responses to risks, and to design risk management and communication strategies for coping with public demands and concerns. From a normative perspective, knowledge about individual perceptions of risk cannot be translated directly into policies (Renn 1990). If perceptions are partially based on biases or ignorance, it does not seem wise to use them as yardsticks for risk reduction. In addition, risk perceptions vary among individuals and groups. Whose perceptions should be used to make decisions on risk? At the same time, however, these perceptions reflect the real concerns of people and include the undesirable effects that "technical" analyses of risk often miss. Facing this dilemma, in which way can risk perception studies contribute to improving risk policies (cf. Fischhoff 1985)? Pertinent benefits may include:

- they can identify and explain public concerns associated with the risk source;
- they can elucidate the context of the risk-taking situation;

- they can enhance understanding of controversies about risk evaluation;
- they can identify cultural meanings and associations linked with special risk arenas;
- based on this knowledge, they can help to articulate objectives of risk policies in addition to risk minimization, such as fairness, procedural equity, and institutional trust;
- they can indicate how to design procedures or policies which incorporate these cultural values into the decision making process;
- they can help to design programs for participation and joint decision making;
- they can provide criteria for evaluating risk management performance and organizational structures for monitoring and controlling risks.

Social-science research on risk perception is useful for risk management in many ways. Firstly, it alerts risk managers to the concerns that people associate with different risks. Based on this information, managers can devote at least parts of their budget with respect to time and resources to those areas that reflect public concerns, in addition to those areas that are required by their mandate or expert judgments. Secondly, it provides risk managers with the necessary information about what individuals and different groups think about a risk and what they feel about it. Both the cognitive and the affective components of risk perception need to be addressed when communicating to the public in general or specific groups in particular. The core argument here is that risk communication, an indispensable component of risk management, cannot be effective without a comprehensive understanding of how people perceive and evaluate risks, and why risk perception varies so much within a society (Fischhoff, Bostrom and Quadrel 1993; Renn 1992a; Rohrmann 1991).

Yet the inclusion of risk perception studies for normative use in policy making faces two major drawbacks. First, the advice of social scientists will vary considerably depending on the perspective that the scientists prefer. Second, unlike the technical or mathematical understanding of risk, the social science concepts offer no common denominator for measuring cultural or social acceptability (Douglas 1985). What constitutes a major risk for one group, may be perfectly in line with the considerations of another group. Who is going to decide which risk perception and evaluation has more validity than an alternative one?

There is also no impartial referee available to judge the appropriateness of risk perceptions. Science may help to determine the magnitude of the risk by providing data on the probability, severity and time frame of particular consequences. This information in itself, however, is not sufficient to make decisions about the acceptability of risks. The only viable resolution of these conflicts in democratic societies is by initiating a discourse among the major parties involved in the decision making process or affected by the decision outcomes (Habermas 1971). Such a dialogue can be organized in form of advisory committees, citizen panels, formal hearings, and others (Baughman 1995; Fiorino 1989; Renn et al. 1993). Risk communication exercised as forums of public discourse must therefore be a crucial element of any risk management strategy.

Research on the risk communication process (including some of the empirical studies of this volume) indicate the need for trust and credibility between the communicators and their audience. They also reveal the continuous trend towards distrust and suspicion on the side of those who are supposed to bear the risks. The credibility of a communication source is closely linked to the perceived past performance record and its openness for public demands. The more institutions comply with the expectations of the public, the more confidence people have in these institutions and the more trust they assign to their messages. Communication efforts alone may be successful to correct excessive aspirations or to mitigate misperceptions of the actual performance record, but it is more than unlikely that communication can compensate for poor performance.

Furthermore, in a climate of general distrust toward social organizations, it is helpful to accept countervailing powers and public control and to provide public access to all relevant information. On the basis of these structural opportunities for public involvement and control, specific communication programs can be designed that include elements of successful information and possibly education and persuasion.

Risk communication - whether organized as providing information to the public, as a mutual learning process, or as an attempt to reconcile conflicts about risks - is therefore a necessary step towards bridging the gap between the conclusions drawn from quantitative risk analysis and inferences based on risk perception. However, the goal of risk communication should not be to induce people to accept whatever the communicator thinks is best for them. The ideal communication program envisions an active citizen who processes all the available information to form a well-balanced judgment in accordance with the factual evidence, the arguments of all sides, and his/her own interests and needs. The ultimate goal of risk communication is to reconcile expertise, interests, and public preferences across the cultures within a society and between societies. This goal cannot be achieved without accepting risk perception as a legitimate expression of people's view of the world and their vision of a "good" life.

5 SOME CONCLUSIONS FOR FURTHER RISK RESEARCH

While research on cross-cultural risk perception is certainly a thriving field, the knowledge gained so far is still incomplete and the validity concerns expressed earlier in this chapter are salient as well. From a methodological viewpoint, at least three issues deserve serious attention:

- Firstly, as the empirical basis of most studies is quite small, larger sample sizes and more broadly defined sampling techniques are suggested in order to increase generalizability. Furthermore, the sampling rationale should reflect the existing knowledge about psychological, sociological and cultural factors that shape risk judgments and evaluations.
- Secondly, risk perception research using individual data tends to ignore cultural and societal factors that are hard to define as variables for person-centered data sets. Consequently, crucial context effects are excluded. At the same time, research investigating processes at a group or society level might underestimate the relevance of individual characteristics and inter-individual differences. Thus, respective models would be difficult to apply to individual risk appraisal and behavior. Therefore, conceptualizations that operate on both an individual

and an aggregate level (cf. "individualism" versus "centralism", Krimsky 1992) would be most valuable when designing future studies.

- Thirdly, the statistical analysis of risk perception data - while certainly sophisticated in the studies collected in this volume - might be further refined in future research. Useful (but rather seldom used) methods include: non-metric multidimensional scaling, multi-mode factor analysis, structural (causal) modeling, multi-level impact analysis, content analysis or influence diagrams for qualitative responses (see e.g., Burns & Clemen 1993, Lappe 1991, Rohrmann 1991, Schuetz in press, Slovic 1992 for a discussion and examples). Given the influence methodological factors may exert on the nature of a study's results, it would be valuable to identify and investigate systematically the substantive implications of the methodology chosen by each respective researcher.

Of course, poor research design and bad sampling methods cannot be "healed" through sophisticated statistical procedures. However, the multivariate and multi-level nature of risk perception processes (as depicted in fig. 2) should be reflected appropriately in the chosen statistical approach.

Regarding relevant *substantive* issues for future research on cross-cultural risk perception, the studies presented in this volume as well as the review in the introductory chapter point at quite a number of open questions deserving continuous research (cf. also Rohrmann 1999):

- *Coverage of countries and cultures:* Most studies were conducted in Europe, North America and more recently in (East-)Asia (cf. the documentation in Rohrmann 1999), while almost nothing is known yet with respect to South-American or (Black-)African or Arabic/Islamic cultures (at least not in terms of internationally published research).
- *Demographic factors:* Most risk perception studies are restricted to student samples, while only a few look at the population as a whole or specific societal groups. Findings on the effects of gender (e.g., Flynn, Slovic & Mertz 1994, Savage 1993), occupation (e.g., Bellrose & Pilisuk 1991) or ethnicity (e.g., Vaughn & Nordenstam 1991) are still incoherent. Clearly, demographic and ethnic variables (and their interaction with cultural orientations) deserve further investigation.
- *Generation effects:* In many cultures adolescents develop their own subcultures. One of the premises is that younger people are more prepared to take risks (Bell & Bell 1993). The risk perception of adolescents (Drottz-Sjöberg & Sjöberg 1991, Greening & Dollinger 1992) has received only minor attention so far.
- *Influence of experience:* In most investigations the respondents have little or no personal exposure to and knowledge of the risk sources under study (although some research has been conducted with people such as divers, firefighters or smokers). Differences in risk perception between exposed and unexposed individuals need systematic examination.
- *Risk attitudes:* A related issue is the role of risk attitudes such as sensation seeking, risk propensity or proneness, and pursuit of adventure on risk perception processes (Bromiley & Curley 1992, Farley 1991, Keinan et al. 1984, Zuckerman 1990). The few available findings (e.g., Dahlbaeck 1996, Horvath & Zuckerman 1992, Rohrmann & Loquet in prep., Sitkin &

Weingart 1995, Weber & Hsee in press) do not give a coherent picture. Risk attitudes and their influence on the evaluation of risks is likely to vary considerably across cultures.

- *Judgment-behavior link:* Risk perception research deals with judgments, that is, verbal reactions to (more or less hypothetical) risk situations. How such cognitions (or emotions) about risks are related to actual behavior, e.g., risk-avoiding activities, job change, protests against risky technologies, and so on, has not been investigated thoroughly (see, e.g., Horvath & Zuckerman 1992, Schoemaker 1993, Trimpop 1994, Weinstein 1988). Again, discrepancies between different cultures are to be expected.
- *Theoretical foundation:* Causes and effects in the cognitive process underlying risk evaluations are not yet sufficiently clarified. A related shortcoming is a lack of comprehensive theoretical framework. A number of existing theories are (at least partially) relevant for risk perception - see, e.g., Kasperson et al. 1988 or Renn et al. 1992 on social risk amplification, Lopes' motivational conceptualization (1988, 1993), Wilde's risk homeostasis theory (1988), Trimpop's risk motivation theory (1994), Weinstein's precaution adoption process (1988). Their application in empirical risk perception studies are rare (Krimsky 1992), and there have been no attempts to test the cross-cultural validity of such conceptualizations yet.
- *"Psychometric" versus "cultural-theory" approach:* These two approaches to cross-cultural risk perception research have led to heterogeneous conceptualizations and models, they differ in empirical validity, and cross-validation is rare (cf., e.g., Rohrmann & Renn, this vol., or - critically - Sjöberg 1997). More 'bridging' research (as attempted by, e.g., Marris et al. 1996, Slovic et al., this vol., or Steg 1998) would be beneficial.
- *New integrated conceptualizations of risk:* Finally, studies on risk perception could be further utilized to refine the conceptualization of "risk" (Drottz-Sjöberg 1991, Renn 1992b, Rohrmann 1998, Smithson 1991) and to construct a conceptual bridge between quantitative and qualitative definitions.

Given the wide range of available methods, as well as the awareness of what we know and - more relevant here - what we do not yet understand sufficiently, a 'second generation' of sophisticated risk research seems worthwhile. Cross-cultural studies - both empirical and theoretical/conceptual ones - should be an exciting field for such an enterprise.

As much as we would like to see more research on cross-cultural risk perception, the studies included in this volume as well as the existing literature on this subject already indicate the usefulness and practicability of the insights gained so far. The need for more research cannot justify not acting on what we already know. The results of risk perception studies have immanent value for making prudent risk management decisions, and for enhancing the capability of society and its institutions to cope with risks in accordance with the preferences, values, and visions of their citizens. In a world characterized by globalization and international cooperation, cross-cultural research is an essential tool to facilitate the understanding between nations and cultures, to ensure that the risks that people fear in different social and cultural contexts are met with the institutional means to manage and to control them appropriately.

6 REFERENCES

Allen, F.W. (1987). Towards a holistic appreciation of risk: the challenge for communicators and policymakers. Science, Technology, and Human Values, 12, 138-143.

Baughman, M. (1995). Mediation. In: O. Renn, T. Webler, & P. Wiedemann (Eds.), Fairness and competence in citizen participation: Evaluating models for environmental discourse. Dordrecht & Boston: Kluwer.

Bell, N. J., & Bell, R. W. (1993). Adolescent risk taking. London: Sage.

Bellrose, C., & Pilisuk, M. (1991). Vocational risk tolerance and perceptions of occupational hazards. Basic and Applied Social Psychology, 12, 303-323.

Bradbury, J.A. (1989). The policy implications of differing concepts of risk. Science, Technology, and Human Values, 14 (No.4), 380-399.

Bromiley, P., & Curley, S. P. (1992). Individual differences in risk taking. In J. F. Yates (Ed.), Risk taking behavior (pp. 87-133). Chichester: John Wiley & Sons.

Burns, W.J. & Clemen, R.T. (1993). Covariance structure models and influence diagrams. Management Science, 39, 816-834.

Clarke, L. (1989). Acceptable risk? Making choices in a toxic environment. Berkeley: University of California.

Coleman, C. (1993). The influence of mass media and interpersonal communication on societal and personal risk judgments. Communication Research, 20, 611-628.

Cvetkovich, G., & Earle, T. C. (Eds.) (1991). Risk and culture. Journal of Cross-cultural Psychology, special volume.

Dahlbaeck, O. (1996). Personality and risk taking. Personality and Individual Differences, 11, 1235-1242.

Douglas, M. (1985). Risk acceptability according to the social sciences. New York: Russell Sage Foundation.

Douglas, M. & Wildavsky, A. (1982). Risk and culture. Berkeley: University of California Press.

Drottz-Sjöberg, B. (1991). Risk perceptions and definitions of risk. In B. Drottz-Sjöberg (Ed.), Perceptions of Risk (pp. 163-214). Stockholm: Center for Risk Research.

Drottz-Sjöberg, B., & Sjöberg, L. (1991). Adolescents attitudes to nuclear power and radioactive wastes. In B. Drottz-Sjöberg (Ed.), Perception of Risk (pp. 109-135). Stockholm: Center for Risk Research.

Earle, T. C., & Cvetkovich, G. T. (1995). Social trust: Toward a cosmopolitan society. Westport: Praeger.

Farley, F. (1991). The type-T personality. In L. P. Lipsitt & L. L. Mitnick (Eds.), Self-regulatory behavior and risk-taking: Causes and consequences (pp. 371-382). Norwood/NJ: Ablex.

Fiorino, D. J. (1989). Technical and democratic values in risk analysis. Risk Analysis, 9, 293-299.

Fischhoff, B. (1985). Managing risk perceptions. Issues in Science and Technology, 2, 83-96.

Fischhoff, B., Bostrom, A., & Quadrel, M. J. (1993). Risk perception and communication. Annual Review of Public Health, 14, 183-203.

Freudenburg, W.R. (1992). Heuristics, biases, and the not-so-general publics: Expertise and error in the assessment of risks. In S. Krimsky and D. Golding (Eds.), Social theories of risk (pp. 229-250). Westport: Praeger.

Freudenburg, W.R. (1993). Risk and recreancy: Weber, the division of labor, and the rationality of risk perceptions. Social Forces, 71, 909-932.

Flynn, J., Slovic, P., & Mertz, C. K. (1994). Gender, race, and perception of environmental health risks. Risk Analysis, 14, 1101-1108.

Giddens, A. (1990). The consequences of modernity. Stanford: Stanford University Press.

Gould, L.C.; Gardner, G.Y.; DeLuca, D.R.; Tieman, A.; Doob, L.W. & Stolwijk, J.A.J. (1988). Perceptions of technological risk and benefits. New York: Russell Sage Foundation.

Greening, L., & Dollinger, S. J. (1992). Adolescent's perceptions of lightning and tornado risks. Journal of Applied Social Psychology, 22, 755-762.

Habermas J. (1971). Toward a rational society. London: Heinemann.

Hampel, J. & Pfenning, U. (1998). Biotechnology and public perception of technology. The German case. Working Report of the Center of Technology Assessment in Baden-Württemberg. Stuttgart: TA-Center.

Horvath, P., & Zuckerman, M. (1992). Sensation seeking, risk appraisal and behavior. Personality and Individual Differences, 14, 41-53.

Jasanoff, S. (1998). The political science of risk perception. Reliability Engineering & Systems Safety, 59 (1), 91-99.

Kahneman, D. & Tversky, A. (1974). Judgment under uncertainty. Heuristics and biases. Science, 185, 1124-1131.

Kahneman, D. & Tversky, A. (1979). Prospect theory: An analysis of decision under risk. Econometrica, 47 (No. 2), 263-291.

Kals, E. (1996). Are proenvironmental commitments motivated by health concerns or by perceived justice? In: L. Montada & M. Lerner (Eds.), Current societal concerns about justice (pp. 231-258). New York: Plenum

Kals, E., Schumacher, D. & Montada, L. (1999). Emotional affinity toward nature as motivational basis to protect nature. Environment & Behavior, 31 (2), 178-202

Kasperson R., Golding, D. & Tuler, S. (1992). Social distrust as a factor in siting hazardous facilities and communicating risks. Journal of Social Issues, 48, 161-187

Kasperson, R.E. & Kasperson, J.X. (1983). Determining the acceptability of risk: Ethical and policy issues. In J.T. Rogers & D.V. Bates (Eds.), Assessment and perception of risk to human health (pp. 135-155). Conference Proceedings. Ottawa: Royal Society of Canada.

Kasperson, R.E, Renn, O., Slovic, P., Brown, H.S., Emel, J., Goble. R., Kasperson, J.X. & Ratick, S. (1988), The social amplification of risk: a conceptual framework. Risk Analysis, 8, 177-187.

Keinan, G., Meir, E., & Gome-Nemirovsky, T. (1984). Measurement of risk takers' personality. Psychological Reports, 55, 163-167.

Krimsky, S. (1992). The role of theory in risk studies. In S. Krimsky & D. Golding (Eds.), Social theories of risk (pp. 3-22). Westport: Praeger.

Kunreuther, H. & Slovic, P. (1996). Science, values, and risk. Annals of the American Academy of Political and Social Science, Special Issue, H. Kunreuther & P. Slovic (Eds.), Challenges in risk assessment and risk management (pp. 116-125). Thousand Oaks: Sage.

Lantermann, E,-D., Döring-Seipel, E. & Schirma, P. (1992). Werte, Gefühle und Unbestimmtheit: Kognitivemotionale Wechselwirkungen im Umgang mit einem ökologischen System. In:

K. Pawlik & K.H. Stapf (Eds.), Umwelt und Verhalten. Perspektiven und Ergebnisse ökopsychologischer Forschung (pp. 129-144). Bern: Huber

Lappe, H. (1991). Perceptions of drug risk: individual and public responses to drugs. Ph.Diss., Berlin: Technical University of Berlin.

Lichtenberg, J., & MacLean, D. (1988). The role of media in risk communication. In H. Jungermann, R. E. Kasperson, & P. M. Wiedemann (Eds.), Risk communication (pp. 33-49). Proceedings of the international workshop on risk communication. Juelich: Research Center - MUT.

Linnerooth-Bayer, J. & Fitzgerald, K.B. (1996). Conflicting views on fair siting processes: Evidence from Austria and the U.S. Risk: Health, Safety & Environment, 7 (2), 119-134

Lopes, L. L. (1988). Between hope and fear: The psychology of risk. Advances in Experimental Social Psychology, 20, 255-295.

Lopes, L. L. (1993). On modeling risky choice: Why reasons matter. In O. Huber, J. Mumpower, J. V. Plight, & P. Koele (Eds.), Current themes in psychological decision research . Amsterdam: Elsevier.

Luhmann N. (1990). Technology, environment, and social risk: a systems perspective. Industrial Crisis Quarterly, 4, 223-231.

Marris, C., Langford, I., & O'Riordan, T. (1996). Integrating sociological and psychological approaches to public perceptions on environmental risks. Research Report. Norwich: University of East Anglia.

MacLean, D. (1986). Social values and the distribution of risk. In D. MacLean (Ed.), Values at risk (pp. 75-93). Totowa: Rowman & Allanheld.

Peters, H.P. (1986). Public opinion as a channel of communication between science and other parts of society. Manuscript for the 11. World Congress of Sociology, August 18-22, 1986. New Delhi: Research Center Juelich- MUT.

Peters, H.P. (1994): Risikokommunikation in den Medien. In: K. Merten; S.J. Schmidt & S. Weichenberger (Eds.), Die Wirklichkeit der Medien. Eine Einführung in die Kommunikationswissenschaft (pp. 329-351). Opladen: Westdeutscher Verlag

Rayner, S. (1984). Disagreeing about risk: The institutional cultures of risk management and planning for future generations. In S.G. Hadden (Ed.), Risk analysis, institutions, and public policy (pp. 150-17). Port Washington: Associated Faculty Press.

Rayner, S. (1987). Risk and relativism in science for policy. In V. T. Covello und B.B. Johnson (Eds.), The social and cultural construction of risk (pp. 5-23). Dordrecht: Reidel.

Rayner, S. (1990). Risk in cultural perspective: Acting under uncertainty. Norwell: Klewer.

Rayner, S. & Cantor, R. (1987). How fair is safe enough? The cultural approach to societal technology choice. Risk Analysis, 7, 3-13.

Renn, O. (1990). Risk perception and risk management: A review. Risk Abstracts, 7 (No. 1), 1-9 (Part 1) and 7 (No. 2), 1-9 (Part 2).

Renn, O. (1992a). Risk communication: towards a rational discourse with the public. Journal of Hazardous Materials, 29, 465-519.

Renn O (1992b). The social arena concept of risk debates. In S. Krimsky and D. Golding (Eds.), Social theories of risk (pp. 170-197). Westport: Praeger.

Renn O (1995). Style of using scientific expertise: a comparative framework. Science and Public Policy, 22, 147-156.

Renn, O., Burns, W. J., Kasperson, J. X., Kasperson, R. E., & Slovic, P. (1992). The social amplification of risk: Theoretical foundations and empirical applications. Journal of Social Issues, 48, 137-160.

Renn, O., Webler, Th., Rakel, H., Dienel, P. C., & Johnson, B. B. (1993). Public participation in the decision making: a three-step procedure. Policy Sciences, 26, 189-214.

Renn, O. & Zwick, M.M. (1997): Risiko- und Technikakzeptanz. Berlin et. al.: Springer.

Rohrmann, B. (1991). Akteure der Risiko-Kommunikation. In H. Jungermann, B. Rohrmann, & P. M. Wiedemann (Eds.), Risikokontroversen - Konzepte, Konflikte, Kommunikation (pp. 355-371). Berlin et.al.: Springer.

Rohrmann, B. (1998). The risk notion - epistemological and empirical considerations. In R. Melchers (Ed.), Integrative risk assessment II. Rotterdam: Balkema.

Rohrmann, B. (1999). Risk perception research: Review and documentation (Update, Studies in Risk Communication vol. 69). Juelich: Research Center. (Available *on the WWW; URL* = http://www.kfa-juelich.de/mut/hefte/heft_69.pdf).

Rohrmann, B. & Loquet, T. (in prep). Attitudes toward risk taking and risk perception: Findings from five studies (Research report). Melbourne: University of Melbourne, Dept. of Psychology.

Savage, I. (1993). Demographic influences on risk perceptions. Risk Analysis, 13, 413-420.

Schoemaker, P. J. H. (1993). Determinants of risk-taking: behavioral and economic views. Journal of Risk and Uncertainty, 49-73.

Schuetz, H. (in prep.). Die Struktur von Risikowahrnehmung - Anwendung von Strukturgleichungsmodellen. Juelich: Research Center Juelich - MUT.

Schwarz, M. & Thompson, M. (1990). Divided we stand: Redefining politics, technology, and social choice. Philadelphia: University of Pennsylvania Press.

Short, J.F. (1984). The social fabric of risk: Toward the social transformation of risk analysis. American Sociological Review, 49, 711-725.

Short, J.F. (1989). On defining, describing, and explaining elephants (and reactions to them): Hazards, disasters, and risk analysis. Mass Emergencies and Disasters, 7 (No.3), 397-418.

Short, J.F & Clarke, L. (1992). Social organization and risk. In J.F. Short & L. Clarke (Eds.), Organizations, uncertainties, and risk (pp. 309-321). Boulder: Westview.

Shubik, M. (1991). Risk, society, politicians, scientists, and people. In M. Shubik (Ed.), Risk, organizations, and society (pp. 7-30) Dordrecht: Kluwer.

Sitkin, S. B., & Weingart, L. R. (1995). Determinants of risky decision-making Behavior: A test of the mediating role of risk perceptions and propensity. Academy of Management Journal, 38, 1573-1592.

Sjöberg, L. (1997). Explaining risk perception: An empirical evaluation of cultural theory. Risk Decision and Policy, 113-130.

Slovic, P., Fischhoff, B., & Lichtenstein, S. (1985). Characterizing perceived risk. In R. W. Kates, C. Hohenemser, & J. X. Kasperson (Eds.), Perilous progress: managing the hazards of technology (pp. 91-125). Boulder: Westview.

Slovic, P. (1992). Perception of risk: Reflections on the psychometric paradigm. In D. Golding & S. Krimsky (Eds.), Theories of risk (pp. 117-152). London: Praeger.

Slovic, P., Layman, M., & Flynn, J. (1991). Risk perception, trust, and nuclear waste: Lessons from Yucca Mountain. Environment, 33, 6-11, 28-30.

Smithson, M. (1991). The changing nature of ignorance. In J. Handmer, B. Dutton, B. Guerin, & M. Smithson (Eds.), New perspectives on uncertainty and risk (pp. 5-38). Canberra: ANU/ CRES.

Steg, L. (1998). Myths of nature and risk perception: A psychological contribution to cultural theory. Paper presented at the ICAP 98.

Trimpop, R. M. (1994). The psychology of risk taking behavior. Amsterdam: New Holland.

Urban, D. (1998): Technikeinstellungen: gibt es die überhaupt? Eine Längsschnittanalyse von Bewertungen der Gentechnik. Working Report of the Institute for Social Sciences, No.2. Stuttgart: University of Stuttgart

Vaughn, E., & Nordenstam, B. (1991). Perceptions of environmental risk among ethnically diverse groups in the United States. Journal of Cross Cultural Psychology, 22, 29-60.

Weber, E. U., & Hsee, C. (1999 in press). Cross-cultural differences in risk perception, but cross-cultural similarities in attitudes towards perceived risk. Management Science

Weinstein, N. (1988). The precaution adoption process. Health Psychology, 355-386.

Wilde, G. J. S. (1988). Risk homeostasis theory. Ergonomics, 31, 491-506.

Wilkins, L. (1991). The risks outside and the pictures in our heads: Connecting the news to people and politics. In J. Handmer, B. Dutton, B. Guerin, & M. Smithson (Eds.), New perspectives on uncertainty and risk (pp. 133-160). Canberra: ANU/ CRES.

Zuckerman, M. (1990). The psychology of sensation seeking. Journal of Personality, 58, 313-345.

Zwick, M.M. & Renn, O. (1998). Wahrnehmung und Bewertung von Technik in Baden-Württemberg. Report Center of Technology Assessment in Baden-Württemberg. Stuttgart: Center of TA.

LIST OF EDITORS AND CONTRIBUTORS

ORTWIN RENN is chair of the Board of Directors at the Center of Technology Assessment in Baden-Württemberg and is Professor for Environmental Sociology at the University of Stuttgart. His professional career included a six year residency as associate professor of environment, technology and society at Clark University (Mass., USA) as well as a visiting professorship at the Swiss Institute of Technology in Zürich (Switzerland). Renn received a MA degree in Sociology and Economics (European Diploma) and a doctorate degree in Social Psychology, both from the University of Cologne (School of Economics and Social Sciences). His current research includes risk perception and communication, regional concepts of sustainable development and citizen participation in risk management. Renn has published 14 books and more than 80 articles in journals or chapters in books. Among others, he serves on the German Government's Advisory Council for Global Environmental Change, on the Future Commission of the state of Baden-Württemberg and the Collegiate of the European Academy of Scientific and Technological Change. Renn was president of the Society for Risk Analysis Europe from 1996 to 1997.

Contact address:

Ortwin Renn, Center of Technology Assessment in Baden-Württemberg,
Industriestrasse 5, D-70565 Stuttgart, Germany
Phone: +49-711-9063 160, Fax: +49-711-9063 175
E-mail: ortwin.renn@ta-akademie.de
Website: http://www.ta-akademie.de

BERND ROHRMANN is Senior Lecturer at the University of Melbourne/Australia and also Adjunct Professor at the University of Mannheim/Germany.

PhD 1974 at Hamburg University (Social Science Faculty); 1983 Habilitation in Psychology at Mannheim University. Since 1992 living and working in Australia. Main research areas include: applied social research, environmental psychology, and research methodology. Special substantive interests: risk perception/communication/ management; impacts of environmental stressors (e.g., noise); hazard appraisal and disaster preparedness; residential choice and satisfaction; decision processes and decision-aiding technologies; teaching quality. Methodological interests: response scales, survey methodology, evaluation research, and structural models. Also extensive work as consultant with governmental agencies, courts and companies. Publication of about 120 articles/ reports/chapters/books.

Contact address:

Bernd Rohrmann, Dept. of Psychology, Univ. of Melbourne, Vic 3052, Australia
Phone: +61-3-9344 6349, Fax: +61-3-9347 6618
E-Mail: b.rohrmann@psych.unimelb.edu.au
Website: http://www.psych.unimelb.edu.au/staff/acad/rohrmann.html

PAUL SLOVIC is president of Decision Research and a professor of psychology at the University of Oregon. He studies human judgment, decision making, and risk analysis. Dr. Slovic received a B.A. degree from Stanford University, an M.A. and Ph.D. degree from the University of Michigan, and an honorary doctorate from the Stockholm School of Economics. He is past president of the Society for Risk Analysis and in 1991 received its Distinguished Contribution Award. In 1993, Dr. Slovic received the Distinguished Scientific Contribution Award from the American Psychological Association, and in 1995 he received the Outstanding Contribution to Science Award from the Oregon Academy of Science.

Contact address:
Paul Slovic, Decision Research, 1201 Oak Street, Eugene, OR 97 401-3575, USA
Phone: +1-541-485 2400, Fax: +1-541-485 2403
E-Mail: pslovic@oregon.uoregon.edu
Website: http://www.decisionresearch.org/personnel.html

JAMES FLYNN is a Senior Research Associate at Decision Research in Eugene, Oregon. He studies public, community, and institutional responses to science and technology with an emphasis on risks to human health and the environment. His research is in the areas of risk communication, risk management, and public policy. His studies have focused on nuclear technologies including nuclear power, radioactive waste, transportation of nuclear materials, and responses to the nuclear weapons complex. He has published more than 40 articles and has been involved in more than 50 studies over the last decade, many of them based upon survey data. He is a member of the American Association for Public Opinion Research, the Society for Risk Analysis and serves on the Board of Directors for the Risk Assessment and Policy Association. He graduated from Eastern Washington State University and completed his graduate work (MA; Ph.D.) at the University of Washington.

Contact address:
James Flynn, Decision Research, 1201 Oak Street, Eugene, OR 97 401-3575, USA

C. K. MERTZ has been the data analyst and manager for Decision Research since 1991. She analyzes experimental and survey data in support of the organization's ongoing research in judgment and decision making. She is experienced in multivariate statistical methods including covariance structural modeling. In addition, she has considerable experience in socioeconomic impact analysis and in community planning in rural areas. She was part of the management and administration team for the socioeconomic impact studies of the high-level radioactive waste repository proposed for Yucca Mountain, Nevada. This study addressed the full-range of socioeconomic impacts of the proposed high-level nuclear waste repository on the counties and communities in southern Nevada, including economic-demographic, community facilities and services, fiscal, intergovernmental relations, social-cultural, and risk effects. Ms. Mertz' experience

with EIS preparation includes several socioeconomic impact assessments of proposed oil shale developments in western Colorado prepared for the U.S. Army Corps of Engineers and the Bureau of Land Management. Ms. Mertz received her B.A. from Oklahoma State University in 1973 and a Masters in Regional and Community Planning from Kansas State University in 1981.
Contact address:
C. K. Mertz, Decision Research, 1201 Oak Street, Eugene, OR 97 401-3575, USA

MARC POUMADÈRE is Associate Professor of Risk Management and Communication at the Ecole Normale Supérieure of Cachan, affiliated with the Research Group on Risk, Information and Decision (GRID). Fields of research include:

- safety practices and culture, and the human impact of new or complex technologies, in contexts such as nuclear power plants and military systems;
- the nature and dynamics of the French public's involvement in risk matters, at both the national and international levels;
- the consequences upon individual and organizational performances of decreased trust in the larger population toward nuclear energy;
- social factors contributing to the intensification or attenuation of the perception of a risk event within a community's experience.

Marc Poumadère is past President of the Society for Risk Analysis-Europe, and serves on governmental and European research and advisory committees.
Contact Address:
Marc Poumadère, Ecole Normale Supérieure de Cachan, GRID, 61, Avenue du Président Wilson, 94 235 Cachan Cedex, FRANCE
Phone +33-1-4740 0990, Fax +33-1-4740 8258
E-mail: cnrsgrid@grid.ens-cachan.fr

CLAIRE MAYS is a social psychologist (Harvard, AB '81; U. Paris VII, DESS '85) who has lived and worked in France since 1982. Her acquaintance with the nuclear field began in 1984 when Institut Symlog, a private research consultancy, was called in by managers of a French nuclear power plant to create a participative program of improving worker safety, taking into account group behavioral norms and safety culture. This intervention sparked interest in the perception of risks, which seems to differ from group to group in the workplace as in society, and, in how groups communicate about risks.
The radioactive waste management issue, seen from this perspective of risk communication in society, became a major focus in 1990 with a European Commission (CEC) contract comparing public information practices in Great Britain and in France. Claire Mays is currently appointed (1999-2000) by the United States Academy of Sciences National Research Council to the steering committee of a Board on Radioactive Waste Management initiative.
Other work in the field undertaken by Institut Symlog includes:

- With France's Institut de Sûreté et de Protection Nucléaire: RISKPERCOM, a 5-country

study for the CEC on public perceptions (survey and case study data);
- With University of Surrey and UK-NRPB: SPARPA, a CEC study on Social Psychological Aspects of Radiation Protection after Accidents.

Claire is Editor of the *European Review of Applied Psychology*, a bilingual quarterly. She is the elected Secretary of the Society for Risk Analysis-Europe (1999-2000). A dominant theme in Claire's work is bringing to institutional decision makers, a sense of how individual members of the public may experience risk.

Contact Address:
Claire Mays, Institut Symlog, B. P. 125, 94 232 Cachan Cedex, France
Phone +33-1-4740 0990, Fax +33-1-4740 8258
E-mail: claire.mays@post.harvard.edu

EUGENE A. ROSA is currently the Edward R. Meyer Distinguished Professor of Environmental and Natural Resource Policy in the Thomas S. Foley Institute for Public Policy and Public Service, Professor and Chair of Sociology, Faculty Associate in the Social and Economic Sciences Research Center, and Adjunct Professor in the Program in Environmental Science and Regional Planning, all at Washington State University. He also serves as Secretary of Section K (Social, Political, and Economic Sciences) of the American Association for the Advancement of Science. His research program has focused on environmental topics - particularly energy, technology, and risk issues - with attention to both theoretical and policy concerns. He has investigated the relationship between levels of energy consumption and societal well-being, public opinion about energy problems and policies, factors affecting the adoption of solar technologies and conservation practices, public attitudes toward and acceptance of nuclear power and nuclear policies, and most recently, cross-cultural perceptions of risks. Dr. Rosa is widely published and acclaimed in these areas of research. His current research foci are on the epistemology of risk and on global environmental change.

Contact Address:
Eugene A. Rosa, Department of Sociology, Washington State University, Pullman, WA 99164-4020, USA
Phone: +1-509-335 4621, Fax: +1-509-335 6419
E-mail: rosa@wsu.edu
Website: http://cooley.libarts.wsu.edu/rosa

NORIYUKI MATSUDA. Main fields of research: Risk perception, decision support and analysis.

Contact Address:
Noriyuki Matsuda, University of Tsukuba, Institute of Policy and Planning Sciences, 1-1-1 Tennou-dai, Tsukuba, Ibaraki 305-8573, JAPAN
Phone: +81-298-53 5170, Fax: +81-298-55 3849
E-Mail: mazda@shako.sk.tsukuba.ac.jp
Website: http://www.sk.tsukuba.ac.jp/~mazda/

RANDALL R. KLEINHESSELINK, PhD, is Associate Professor of Psychology at Washington State University Vancouver. He also has a faculty appointment in Environmental Psychology/ Regional Planning. In 1980, he was a visiting scientist at the American Psychological Association in Washington, D.C. In 1986, he was a visiting scientist at the University of Washington Seattle. Dr. Kleinhesselink has received two teaching awards during his career and has also received the Academic Advisor of the Year Award from the Golden Key National Honor Society in 1995. Since transferring from the Pullman campus to the Vancouver campus in 1988, he has served as the Coordinator of the Social Sciences, Coordinator of Psychology, Director of Public Affairs, and Assistant Chair of Psychology. In addition to studying the cross-cultural perception of environmental risks, his research includes studies of selective cognitive attention to information, interpersonal attraction, gender differences, and self-concept/self efficacy. He received his PhD in Social/Personality Psychology from the University of Iowa in 1971.

Contact Address:

Randall R. Kleinhesselink, Assistant Chair of Psychology, Washington State University, Department of Psychology, 14204 NE Salmon Creek Avenue, Vancouver, WA 98 686-9600, USA
Phone +1-360-546 9723
E-Mail: kleinhes@vancouver.wsu.edu
Website: http://www.vancouver.wsu.edu/fac/kleinhes/home.htm

LENNART SJÖBERG is Head of the Center for Risk Research of the Stockholm School of Economics and a Fellow of the Society for Risk Analysis. He is Associate Editor of Journal of Risk Research, and a member of the editorial board of Risk Analysis, An International Journal. Prof. Sjöberg has been working on risk perception and related issues since the 1970´s, and most recently coordinated CEC project RISKPERCOM which involved 5 countries in Western Europe.

Contact Address:

Lennart Sjöberg, Stockholm School of Economics, Center for Risk Research, Box 6501
113 83 Stockholm, Sweden
Phone: +46-8-736 9580, Fax +46-8-30 72 25
E-Mail: pls@hhs.se
Website: http://www.hhs.se/dfr/faculty/Sjoeberg.htm

DANIELA KOLAROVA is a lecturer in Social Psychology and Social Ecology at Sofia University "St. Kl. Ohridski". Last years Dr. Daniela Kolarova participated in several international research projects on risk perception and risk communication. During 1996-1997 she served as Bulgarian principal investigator of the EU funded project Risk Perception, the Environment, and Communication Strategies in the CESAR Project (a project on environmental health, public risk perception and risk communication in six Central and Eastern European Countries).

Contact Address:

Daniela Kolarova, Brjast str.20, Sofia 1164, Bulgaria
Phone/Fax: 359 2 66 43 18
E-mail: daniela@bitex.com

INDEX

P

R

S

T

U

W

Technology, Risk, and Society

An International Series in Risk Analysis

---

1. J.D. Bentkover, V.T. Covello and J. Mumpower (eds.): *Benefits Assessment.* The State of the Art. 1986 ISBN 90-277-2022-3
2. M.W. Merkhofer: *Decision Science and Social Risk Management.* A Comparative Evaluation of Cost-Benefit Analysis, Decision Analysis, and other Formal Decision-Aiding Approaches. 1987 ISBN 90-277-2275-7
3. B.B. Johnson and V.T. Covello (eds.): *The Social and Cultural Construction of Risk.* Essays on Risk Selection and Perception. 1987 ISBN 1-55608-033-6
4. R.E. Kasperson and P.J.M. Stallen (eds.): *Communicating Risks to the Public.* International Perspectives. 1990 ISBN 0-7923-0601-5
5. D.P. McCaffrey: *The Politics of Nuclear Power.* A History of the Shoreham Nuclear Power Plant. 1991 ISBN 0-7923-1035-7
6. M. Waterstone (ed.): *Risk and Society.* The Interaction of Science, Technology and Public Policy. 1991 ISBN 0-7923-1370-4
7. A. Vari and P. Tamas (eds.): *Environment and Democratic Transition.* Policy and Politics in Central and Eastern Europe. 1993 ISBN 0-7923-2365-3
8. A. Vari, P. Reagan-Cirincione and J.L. Mumpower: *LLRW Disposal Facility Siting.* 1994 ISBN 0-7923-2743-8
9. B. Brehmer and N.-E. Sahlin (eds.): *Future Risks and Risk Management.* 1995 ISBN 0-7923-3057-9
10. O. Renn, T. Webler and P. Wiedemann (eds.): *Fairness and Competence in Citizen Participation.* Evaluating Models for Environmental Discourse. 1995 ISBN Hb 0-7923-3517-1; Pb 0-7923-3518-X
11. P.C.R. Gray, R.M. Stern and M. Biocca (eds.): *Communicating About Risks to Environment and Health in Europe.* 1997 ISBN 0-7923-4519-3
12. R.E. Jorissen and P.J.M. Stallen (eds.): *Quantified Societal Risk and Policy Making.* 1998 ISBN 0-7923-4955-5
13. O. Renn and B. Rohrmann (eds.): *Cross-Cultural Risk Perception.* A Survey of Empirical Studies. 2000 ISBN 0-7923-7747-8

Kluwer Academic Publishers – Dordrecht / Boston / London